D0197496

HOW TO SOLVE
MATHEMATICAL PROBLEMS

Wayne A. Wickelgren

DOVER PUBLICATIONS, INC.
New York

Bibliographical Note

This Dover edition, first published in 1995, is an unabridged, slightly corrected republication of the work first published by W. H. Freeman and Company, New York, 1974, under the title *How to Solve Problems: Elements of a Theory of Problems and Problem Solving*.

Library of Congress Cataloging-in-Publication Data

Wickelgren, Wayne A., 1938–
    [How to solve problems]
    How to solve mathematical problems / Wayne A. Wickelgren.
       p.   cm.
    Originally published: San Francisco : W.H. Freeman, 1974, in series: A series of books in psychology.
    Includes bibliographical references and index.
    ISBN 0-486-28433-6 (pbk.)
    1. Mathematics—Problems, exercises, etc. 2. Problem solving. I. Title.
QA43.W52   1995
511.3—dc20   94-38619
                                                              CIP

Manufactured in the United States of America
Dover Publications, Inc., 31 East 2nd Street, Mineola, N.Y. 11501

*For as long as I can remember, I have been more
interested in reflecting on what I was doing or
thinking and in thinking about ways to improve my
methods than I have been in the particular things
I was doing or thinking about. This emphasis on
self-analysis and improvement reflects the influence
of my mother and father, Alma and Herman Wickelgren,
to whom this book is dedicated and whose values and
practical principles have contributed so much
to my life.*

Contents

Preface

In the mathematics and science courses I took in college, I was enormously irritated by the hundreds of hours that I wasted staring at problems without any good idea about what approach to try next in attempting to solve them. I thought at the time that there was no educational value in those "blank" minutes, and I see no value in them today. The general problem-solving methods described in this book virtually guarantee that you will never again have a blank mind in such circumstances. They should also help you solve many more problems and solve them faster. But whether or not you solve any particular problem, you will always have lots of ideas about ways to attack the problem. Also, the use of general problem-solving methods often indicates the properties of the principles you need to know from the subject matter that the problem is attempting to teach and test. Thus, whether you succeed or fail in solving any particular problem, the effort will be interesting and educational.

The theoretical and practical analyses of problems and problem solving presented here were heavily influenced by advances made over the last 20 years in the fields of artificial intelligence and computer simulation of thought. My greatest intellectual debts are to Allen Newell, Herbert Simon, and George Polya. Newell and Simon's analyses of problems and problem solving constituted my starting point for working in this area, and many of the best ideas in the book are ideas they have already presented in one form or another. Many other good ideas were taken more or less directly from Polya, whose books on mathematical problem solving are a rich source of methods and a stimulus for thought.

My efforts to understand and organize problem-solving methods began in 1959 when, as an undergraduate at Harvard, I first became aware of the pioneering work of Allen Newell, Cliff Shaw, and Herb Simon on the computer simulation of thinking. During graduate school at the University of California, Berkeley, I regarded problem solving as my major research area. I do not think that my experimental studies of human problem solving ever amounted to much. However, I thought at the time (and think today) that my theoretical (mathematical) understanding of problems and problem solving was immeasurably increased and that this greatly enhanced my ability to solve all kinds of mathematical problems. Shortly after coming to MIT as a new faculty member in the Psychology Department, I decided that one contribution I could make to the undergraduates there was to teach them this newly acquired skill of mathematical problem solving. The students enjoyed the course and, more important, reported back to me in later years that they thought that their problem-solving ability in mathematics, science, and engineering courses had been greatly increased by learning these general problem-solving methods. Enrollment in the course went from 20 to 80 in three years, when I stopped giving it because my primary research interest had shifted to human memory. Some years later, after moving to the University of Oregon, I decided that I now had the time to write a book containing all the ideas that I had acquired from others and generated myself concerning problems and problem solving.

The purpose of the book is to improve your ability to solve all kinds of mathematical problems whether in mathematics, science, engineering, business, or purely recreational mathematical problems (puzzles, games, and so on). This book is primarily intended for college students who are currently taking elementary mathematics, science, or engineering courses. However, I hope that students with less mathematical background can read the book and master the methods without an undue degree of additional effort and also that more advanced readers will profit from it without being bored. I believe that almost everyone who solves mathematical problems can profit substantially from learning the general problem-solving methods described here, and I have tried to write in a way that will communicate effectively to all such people. The approach is to define each general problem-solving method and illustrate its application to simple recreational mathematics problems that require no more mathematical background than that possessed by someone with a year of high school algebra and a year of plane geometry. An elementary knowledge of "new mathematics" (sets, relations, functions, probability, and so on) would be helpful, and some of this is briefly taught in Chapter 10.

The solutions to example problems are presented gradually, usually in the form of hints to give the reader more and more chances to go back and solve the problem. This technique is founded on the belief that you will remember best what you discover for yourself. The book aims to guide you to discovering how to apply general problem-solving methods to a rich variety of problems. I believe that if you read this book and try to apply the methods to around 50 or 100 of your own problems, you will improve substantially in problem-solving ability, with consequent benefits in job performance, school grades, and "intelligence" test scores (including SAT college entrance exams, and The Graduate Record Exam).

Finally, I would like to make a negative acknowledgment. This book was written in spite of my four-year-old son, Abraham, and my six-year-old daughter, Ingrid, who are such delightful people that I cannot resist spending vast amounts of time with them.

October 1973   *Wayne A. Wickelgren*

1

Introduction

The purpose of this book is to help you improve your ability to solve mathematical, scientific, and engineering problems. With this in mind, I will describe certain elementary concepts and principles of the theory of problems and problem solving, something we have learned a great deal about since the 1950s, when the advent of computers made possible research on artificial intelligence and computer simulation of human problem solving. I have tried to organize the discussion of these ideas in a simple, logical way that will help you understand, remember, and apply them.

You should be warned, however, that the theory of problem solving is far from being precise enough at present to provide simple cookbook instructions for solving most problems. Partly for this reason and partly for reasons of intrinsic merit, *teaching by example* is the primary approach used in this book. First, a problem-solving method will be discussed theoretically, then it will be applied to a variety of problems, so that you may see how to use the method in actual practice.

To master these methods, it is essential to work through the examples of their application to a variety of problems. Thus, much of the book is devoted to analyzing problems that exemplify the use of different methods. You should pay careful attention to these problems and should not be discouraged if you do not perfectly understand the theoretical discussions. The theory of problem solving will undoubtably help those students with sufficient mathematical background to understand it, but students who lack such a background can compensate by spending greater time on the examples.

SCOPE OF THE BOOK

This book is primarily a practical guide to how to solve a certain class of problems, specifically, what I call *formal problems* or just "problems" (with the adjective *formal* being understood in later contexts). Formal problems include all mathematical problems of either the "to find" or the "to prove" character but do not include problems of defining "mathematically interesting" axiom systems. A student taking mathematics courses will hardly be aware of the practical significance of this exclusion, since defining interesting axiom systems is a problem not typically encountered except in certain areas of basic research in mathematics. Similarly, the problem of constructing a new mathematical theory in any field of science is not a formal problem, as I use the term, and I will not discuss it in this book. However, any other mathematical problem that comes up in any field of science, engineering, or mathematics is a formal problem in the sense of this book.

Problems such as what you should eat for breakfast, whether you should marry x or y, whether you should drop out of school, or how can you get yourself to spend more time studying are not formal problems. These problems are virtually impossible at the present time to turn into formal problems because we have no good ways of restricting our thinking to a specified set of given information and operations (courses of action we might take), nor do we often even know how to specify precisely what our goals are in solving these problems. Understanding formal problems can undoubtedly make some contributions to your thinking in regard to these poorly specified personal problems, but the scope of the present book does not include such problems. Even if it did, it would be extremely difficult to specify any precise methods for solving them.

However, formal problems include a large class of practical problems that people might encounter in the real world, although they usually encounter them as games or puzzles presented by friends or appearing in magazines. A practical problem such as how to build a bridge across a river is a formal problem if, in solving the problem, one is limited to some specified set of materials (givens), operations, and, of course, the goal of getting the bridge built.

In actuality, you might limit yourself in this way for a while and, if no solution emerged, decide to consider the use of some additional materials, if possible. Expanding the set of given materials (by means other than the use of acceptable operations) is not a part of formal problem solving, but often the situation presents certain givens in sufficiently disguised or implicit form that recognition of all the givens is an important part of skill in formal problem solving. That skill will be discussed later.

Practical problems or puzzles of the type we will consider differ from problems in mathematics, science, or engineering in that to pose them requires less background information and training. Thus, puzzle problems are especially suitable as examples of problem-solving methods in this book, because they communicate the workings of the methods most easily to the widest range of readers. For this reason, puzzle problems will constitute a large proportion of the examples used in this book—at least prior to the last chapter.

In principle, it might seem that most important problem-solving methods would be unique to each specialized area of mathematics, science, or engineering, but this is probably not the case. There are many extremely general problem-solving methods, though, to be sure, there are also special methods that can be of use in only a limited range of fields.

It may be quite difficult to learn the special methods and knowledge required in a particular field, but at least such methods and knowledge are the specific object of instruction in courses. By contrast, general problem-solving methods are rarely, if ever, taught, though they are quite helpful in solving problems in every field of mathematics, science, and engineering.

GENERAL VERSUS SPECIAL METHODS

The relation between specific knowledge and methods, on the one hand, and general problem-solving methods, on the other hand, appears to be as follows. When you understand the relevant material and specific methods quite well and already have considerable experience in applying this knowledge to similar problems, then in solving a new problem you use the same specific methods you used before. Considering the methods used in similar problems is a general problem-solving technique. However, in cases where it is obvious that a particular problem is a member of a class of problems you have solved before, you do not need to make explicit, conscious use of the method: simply go ahead and solve the problem, using methods that you have learned to apply to this class. Once you have this level of understanding of the relevant material, general problem-solving methods are of little value in solving the vast majority of homework and examination problems for mathematics, science, and engineering courses.

When problems are more complicated, in the sense of involving more component steps, and are not highly similar to previously solved problems, the use of general problem-solving methods can be a substantial aid in solution. However, such complex problems will be encountered only rarely by the beginning mathematics, science, and engineering students taking courses in high school and college. More important to the immediate needs of such students is the role of general problem-solving methods in simple homework and examination problems where one does *not* completely understand the relevant material and does *not* have considerable experience in solving the relevant class of problems. In such cases, general problem-solving methods serve to guide the student to recognize what relevant background information needs to be understood. For example, when one understands the general problem-solving method of setting subgoals, one can often set particular subgoals that directly indicate what types of specific information are being tested (and thereby taught) by a particular problem. One then knows what sections of the textbook to reread in order to understand the relevant material.

If, however, the book is not available, as in many examination situations, general problem-solving methods provide one with powerful general methods for retrieving from memory the relevant background information. For example, the use of general problem-solving methods can indicate for which quantities one needs a formula and can provide a basis for choosing among different alternative formulas. Frequently, a student may know all the definitions, formulas, and so on, but not have strong associations to this knowledge from the cues present in each type of problem to which this knowledge is relevant.

With experience in solving a variety of problems to which the knowledge is relevant, one will develop strong direct associations between the cues in such problems and this relevant knowledge. However, in the early stages of learning the material, a student will lack such direct associations and will need to use general problem-solving methods to indicate where in one's memory to retrieve relevant information or where in the book to look it up. Assuming this idea is true (and this book aims to convince you it is), mastering general problem-solving methods is important to you both so you can use problems as a learning device and so you can achieve the maximum range of applicability of the knowledge you have stored in mind—on an examination, on a job, or whatever.

The goal of this book is to teach as many of these general problem-solving methods as I know about, so that if you spend the time to master these methods you can more effectively learn the subject matter of your courses. Also, since the ability to use the information given in most mathematics, science, and engineering courses is often primarily the ability to solve problems in these fields, the book aims to increase this ability to use knowledge.

RELATION TO ARTIFICIAL INTELLIGENCE

It should be emphasized that this text is primarily a practical how-to-do-it book in a field where the level of precise (mathematical) formulation is far below what I am sure it will be in the future, perhaps even the near future. Artificial intelligence and computer simulation of human problem solving are currently very active fields of research, and results from some of this work have heavily influenced this book. However, theoretical formulations of problem solving superior to those we currently have will eventually make the present formulation outdated. Nevertheless, the methods described in the present book, however imperfectly, can be of substantial benefit to any student who masters them. When someone has a beautiful mathematical theory of problems and problem solving sometime in the future, then clearer and more effective how-to-do-it books can be written. Meanwhile, it is my hope that this book will help many people to solve problems better than they did before.

APPLYING METHODS TO PROBLEMS

As discussed previously, to master the problem-solving methods described in this book, it is necessary to study the example problems illustrating their use. The problems and solutions analyzed in Chapters 3 to 10 illustrate the use of the methods discussed in the particular chapter. Chapter 11 considers a variety of homework and examination problems for mathematics, science, and engineering courses. Of course, you probably have lots of your own problems to solve in school or work, and you should begin using the methods on these problems immediately. Merely reading this book provides only the beginning concepts necessary to mastering general problem-solving methods. *Practice* in using the methods is essential to achieving a high level of skill.

Everyone who solves problems uses many or all of the methods described in this book, but if you are not an extremely good problem solver, you may be using the methods less effectively or more haphazardly than you could be by more explicit training in the methods. At first, the application of such explicitly taught problem-solving methods involves a rather slow, conscious analysis of each problem.

There is no particular reason to engage in this careful, conscious analysis of a problem when you can immediately get some good ideas on how to solve it. Just go ahead and solve the problem "naturally." However, after you solve it or, even better, while you are solving it, analyze what you are doing. It will greatly deepen your understanding of problem-solving methods, and you might discover new methods or a new application of an old method.

As you get extensive practice in using these problem-solving methods you should become so skilled in their use that the process becomes less conscious and more automatic or natural. This is the way of all skill learning, whether driving a car, playing tennis, or solving mathematical problems.

2

Problem Theory

FOUR SAMPLE PROBLEMS

To illustrate the concepts involved in the theory of problems described in this chapter, we will begin with four sample problems.

Instant Insanity

Instant Insanity is the name of a popular puzzle consisting of four small cubes. Each face of every cube has one of four colors: red (R), blue (B), green (G), or white (W). Each cube has at least one of its six faces with each of the four different colors, but the remaining two faces necessarily must repeat one or two of the colors already used.

The exact configurations of colors on the faces of the cubes are shown in Fig. 2-1. The faces of the cubes in the figure have been cut along the edges and flattened out for easy presentation on the two-dimensional page. (To reconstruct the cube in three-dimensions, one would simply cut out the outlined figure, turn the top flap over on the top and the bottom over on the bottom, and wrap the left side and back around to join up with the right side at the rear of the cube.) For convenience, the faces of one cube in the figure have been labeled *front*, *top*, *bottom*, *back*, *left side*, *right side*. If you think of the front cube

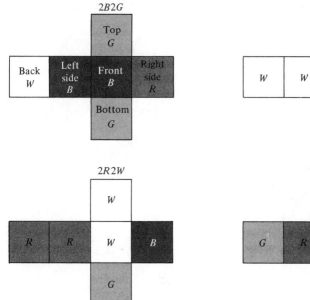

2B2G

	Top G		
Back W	Left side B	Front B	Right side R
	Bottom G		

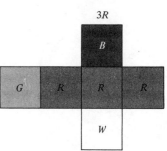

2G2W

2R2W

3R

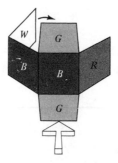

FIGURE 2-1
The six colored faces of each of the four cubes in the
Instant Insanity puzzle. You could cut out each of
the above figures and fold along the edges to make
cubes. In the above figure R = red, B = blue, W =
white, and G = green. The cubes have been given
these "names": $2B2G$, $2G2W$, $2R2W$, and $3R$, which
indicates the colored faces, of which they have more
than one.

as being closest to you (facing you), then mentally constructing the
cube from the two-dimensional drawing should be relatively easy to
do. However, you may wish to buy the puzzle to provide a more concrete and enjoyable representation.

The goal of the puzzle is to arrange the cubes one on top of the other in such a way that they form a stack four cubes high, with each of the four sides having exactly one red cube, one blue cube, one green cube, and one white cube.

Chess Problem

From the board configuration shown in Fig. 2-2 describe a sequence of moves such that white can achieve mate in five moves.

Find Problem from Mechanics

What constant force will cause a mass of 3 kilograms to achieve a speed of 30 meters per second in 6 seconds, starting from rest?

Proof Problem from Modern Algebra

You are given a mathematical system consisting of a set of elements (A, B, C), with two binary operations (call them *addition* and *multiplication*) that combine two elements to give a third element. The system has the following properties: (1) Addition and multiplication are closed; that is, $A + B$ and $A \cdot B$ are members of the original set for all A and B in the set. (2) Multiplication is commutative; that is, AB equals

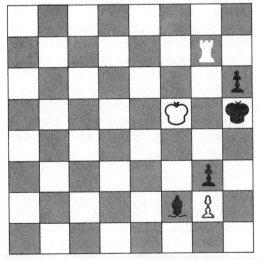

FIGURE 2-2
Part of a famous chess problem. White to achieve mate in five moves.

BA for all *A* and *B* in the set. (3) Equals added to equals are equal; that is, if $A = A'$ and $B = B'$, then $A + B = A' + B'$, for all A, B, A', B' in the set. (4) The left distributive law applies; that is, $C(A + B) = CA + CB$, for all A, B, C in the set. (5) The transitive law also applies; that is, if $A = B$ and $B = C$, then $A = C$. From these given assumptions, you are to prove the right distributive law — that is, that $(A + B)C = AC + BC$, for all A, B, C in the set.

WHAT IS A PROBLEM?

All the formal problems of concern to us can be considered to be composed of three types of information: information concerning *givens* (given expressions), information concerning *operations* that transform one or more expressions into one or more new expressions, and information concerning *goals* (goal expressions). There may be intermediate subgoal expressions mentioned explicitly in the problem, or the problem solver may define these subgoal expressions for himself; but we will assume that there is only one terminal goal per problem. Any problem stated with two or more independent terminal goals could always be viewed as two or more problems with the same givens and operations and different goals.

For convenience and accuracy, I tend to take the more formal view that a problem involves expressions of information rather than actual physical objects. Even in a practical problem stated in terms of physical objects, it is always possible to consider objects or sets of properties of objects as represented by expressions. Indeed, we must have representations in our heads of objects, properties of objects, and operations when we solve practical problems, since we certainly do not have the real objects there. Thus, definitions of problems, solutions, and methods need not make any distinction between practical (concrete) and symbolic (abstract, mathematical). However, when dealing with a practical problem, there is no need to talk of representations or expressions, if the problem is more easily solved without using this more abstract language.

Givens

Givens refer to the set of expressions that we accept as being present in the world of the problem at the onset of work on the problem. Indeed, the givens and the operations together constitute the entire world of the problem at the beginning of work on it. This definition of the givens encompasses expressions representing objects, things, pieces of material, and so on, as well as expressions representing assumptions, definitions, axioms, postulates, facts, and the like.

In some kinds of puzzles the givens consist of the materials. For example, the givens in Instant Insanity are four cubes, with each side of each cube having one of four colors (red, blue, green, or white), as shown in Fig. 2-1.

In the chess problem, the givens are the pieces of each player and their positions on the board plus the information concerning whose move it is. In the particular chess problem shown in Fig. 2-2, the givens are that white has a king, a rook, and a pawn at the positions indicated; that black has a king, a bishop, and two pawns at the positions indicated; and that it is white's move. The implicitly specified given information consists of all the rules of chess, including such information as that the board consists of 64 squares laid out in an 8 × 8 array, the starting positions of the pieces in a game, and that a king can move one square in any direction (horizontally, vertically, or diagonally), that checkmate consists of putting the opponent's king into a position where it would be captured on the next move if it was not moved out of the square it was in and such that all squares that the king could move to would also result in capture.

In the find problem, the givens are the information explicitly stated in the problem plus whatever other mathematical or scientific knowledge is to be implicitly assumed as part of the givens. In the physics problem described above, the explicitly described given information includes the following: the mass of the given object is 3 kilograms, its initial speed is zero, its final speed after 6 seconds of applying a force is 30 meters per second, and the force and mass are constant. Implicitly specified information include Newton's second law that force equals mass times acceleration, and the rules of algebra and possibly calculus (depending upon how one solves the problem).

In a mathematical proof problem, the givens are all the axioms that one is allowed to assume. The givens in the particular proof problem described above are three of the five assumptions: (1) that the system is closed, (2) that multiplication is commutative, and (4) that the left distributive law holds. Assumptions (3), that equals added to equals are equal, and (5), that the transitive law holds (that is, if $A = B$ and $B = C$, then $A = C$) are really rules of inference rather than givens. Rules of inference are operations, discussed below.

Operations

Operations refer to the actions you are allowed to perform on the givens or on expressions derived from the givens by some previous sequence of actions. Other terms for operations include *transforma-tions* and *rules of inference*, though the latter term seems to be appro-priate only for conclusion-drawing problems and not so appropriate for action-oriented problems.

In Instant Insanity, the allowable operations can be conceptualized in a variety of equivalent ways, the simplest of which is just that cubes can be placed on top of one another in a single tower (such that all faces of all cubes are either parallel or perpendicular to one another). In a chess problem, the allowable operations are given by the allow-able moves of each piece on the board of the player whose turn it is to move. In a find problem, the operations are sometimes peculiar to the problem but are often the operations (or rules of inference) of mathematics or logic. In the mechanics problem described at the be-ginning of this chapter, multiplying or dividing both sides of an equa-tion by the same quantity is an allowable operation.

In a proof problem, the operations are those rules of inference that are allowable within the mathematical system in question. For example, in propositional logic, if proposition A is true and if the statement "A implies B" is true, then one may infer that proposition B is true. In the modern-algebra proof problem described at the beginning of the chapter, the two rules of inference that constitute the allowable operations in this problem are property (3), that if $A = A'$ and $B = B'$, then $A + B = A' + B'$, and property (5), that if $A = B$ and $B = C$, then $A = C$. Note that these operations take two input expressions and produce a single new output expression. Also note that, although ad-dition and multiplication are certainly operations within the mathe-matical system described in the proof problem, multiplication and addition are not the operations to be used in solving the problem. Something that is an operation in one problem may be only a part of the given expressions in another problem.

Let me distinguish between *destructive operations*, which produce new expressions by destroying old expressions, and *nondestructive operations*, which produce new expressions to increase the set of existing expressions without destroying any old expressions. In the above examples, Instant Insanity and chess involve destructive opera-tions; algebraic find problems and logical proof problems involve nondestructive operations.

Although many problems allow one to use any allowable operation at any time, some problems place restrictions on the number of times an operation can be used or the conditions under which it can be used. For instance, in chess a pawn first can be moved either one or two squares, but thereafter it can be moved ahead only one square at a time.

Let us adopt the convention that an *operation* refers to a class of *actions*, with the actions being distinguished only by the *operands* — expressions or objects — to which the operation is applied. Assume that a particular operation, F, can be applied to any expression within some set of expressions, $\{x_i\}$. The particular x_i to which we will apply the operation will be called the *operand*. The operation applied to a particular operand, namely, $F(x_i)$, will be called an *action*. Obviously, these definitions of operations, operands, and actions generalize easily to functions of more than one variable — for example, $F(x, y, z)$.

Goals

The *goal* of a problem is a terminal expression one wishes to cause to exist in the world of the problem. There are two types of goals specified in problems: completely specified goal expressions in proof problems and incompletely specified goal expressions in find problems.

For example, consider the problem of finding the value of x, given the expression $4x + 5 = 17$. In this problem, one can regard the goal expression as being of the form $x =$ _____ , where the correct number is to be found in order to fill in the blank in the goal expression. The goal expression in a find problem of this type is incompletely specified. If the goal expression were specified completely — for example, $x = 3$ — then the problem would be a proof problem, with only the sequence of operations to be determined in order to solve the problem. Of course, if one were not guaranteed that the goal expression $x = 3$ was true, then the terminal goal expression should really be considered to be incompletely specified — something like the statement "$x = 3$ is (true or false)."

In Instant Insanity, the goal is incompletely specified. The goal is to get a tower of four cubes arranged in such a way that each of the four rows of sides has one of each of the four colors. However, one is not told exactly what the arrangement of the colors is to be — if one were, it would be a very simple proof problem instead of a rather hard find problem.

In many chess problems, the goal is to checkmate the other player in some small number of moves. This goal is clear, but it is certainly not the same as giving a complete specification of the terminal board position.

Incomplete specification of the goal state does not imply any ambiguity about what constitutes a correct or incorrect solution to the problem, as I shall define the term solution. There may be more than one correct solution to a problem, but all formal problems discussed in this book have the property that a solution is either correct or incorrect, without ambiguity.

One reason for discussing the completeness of specification of the goal is to clearly describe the nature of the difference between find and proof problems. Another reason is to point out that find problems have a terminal or goal expression that is specified (in various ways and to different degrees) in a manner rather similar to the theorem to be proved in a proof problem. It turns out that the degree of similarity in the specification of the goal expression is sufficient to allow most of the same problem-solving methods to be applied to find problems and to proof problems. Working backward from the goal is probably the only general problem-solving method that is used primarily in proof problems and virtually never in find problems. All other methods discussed in this book are frequently used in both find and proof problems. Thus, although the distinction between find and proof problems is perhaps the most familiar distinction between types of problems, it has only moderate significance for problem-solving methods.

Implicit Specification of Givens, Operations, and Goals

Although some problems (for example, some proof problems) explicitly specify all of the givens, operations, and goals, other problems specify them only implicitly. For example, in solving the typical physics problem, all of the assumptions, operations, and previously proved theorems of real-variable and complex-variable mathematics are at one's disposal in working on the problem, though this fact is generally not stated explicitly. Usually, the implicit givens, operations, and goals of a problem are clear to the problem solver, but sometimes they are not.

Incomplete Specification of Givens, Operations and Goals

There are often deliberately incomplete statements of givens, operations, and goals. That is, the problem solver may have some degree of choice among a set of possible given expressions, a set of possible operations, and a set of possible goal expressions. We have already discussed the case where the terminal goal expression is not specified completely, but instead the problem solver has to *find* the correct expression to fill into a blank space in the terminal goal expression. Many find problems, such as the example given earlier of finding $x = \underline{\hspace{1cm}}$, given $4x + 5 = 17$, are equivalent to a problem with a completely specified goal, $4x + 5 = 17$, but with an incompletely specified given, $x = \underline{\hspace{1cm}}$. Equivalences like this obtain where operations are uniquely reversible (that is, where there exist inverse operations for all operations).

In algebra problems — for instance, solving for $x =$ _____ in a cubic equation such as $x^3 + 2x^2 - x - 2 = 0$ — it is probably somewhat better to view the problem as having a completely specified goal expression, $x^3 + 2x^2 - x - 2 = 0$, and an incompletely specified given expression, $x =$ _____, than the reverse. Often you are asked to determine all the values of x that satisfy the equation, which means that you need to know all the values of x from which you could derive the complicated equation. Basically, this is a hypothesis generation (guessing) and testing situation, because the direction of implication (by ordinary arithmetic operations) is from an unknown $x =$ _____ to a known goal, $x^3 + 2x^2 - x - 2 = 0$, not the reverse. There are three values of x that satisfy the equation $x^3 + 2x^2 - x - 2 = 0$, so the latter equation cannot imply three contradictory equations, $x = 1$, $x = -1$, and $x = -2$.

Other examples of problems with incomplete specification of givens or operations include many *construction problems*. Many such problems require one to build something with a range of possible given materials and operations, but there are costs or other restrictions attached to the use of the materials (givens) and operations. The problem solver must select an unordered set of materials and an ordered set of (sequence of) operations that satisfies some constraints specified in the problem and also achieves the goal.

Optimization problems are a natural extension of problems where givens or operations have costs. In an optimization problem, one is supposed to find the way to achieve the goal that minimizes some cost or maximizes some utility.

WHAT IS A PROBLEM STATE?

A *problem state*, the state of the world of a problem, is the set of all the expressions that exist in the world of the problem at a particular time. The problem state can be changed only by applying an operation to one or more expressions existing in the previous problem state to produce one or more new expressions.

In problems that have only nondestructive operations, a problem state consists of all the expressions that have been obtained from the givens up to that moment in working on the problem. In problems that have one or more destructive operations, the problem state includes only the currently existing expressions (those obtained that have not been destroyed). Often problems with destructive operations are considered to have only a single expression representing their state at the current moment, with the operations being able to change that entire state into a new state. In such problems, there is no reason to distinguish between state and expression. The given problem state is the set of all given expressions. When the givens are not specified completely, there are *multiple* possible given states. When the givens are completely specified, there is a *unique* given state. A *goal state* is a state that includes the goal expression. When the goal is not completely specified or when there are nondestructive operations, there are *multiple* possible goal states. When the goal is completely specified and all operations are destructive, there may be a unique goal state.

WHAT IS A SOLUTION?

A *solution* to a problem contains all four of the following parts. (a) Complete specification of the givens; that is, a unique given state from which the goal can be derived via a sequence of allowable operations. (b) Complete specification of the set of operations to be used. (c) Complete specification of the goals. (d) An ordered succession or sequence of problem states, starting with the given state and terminating with a goal state, such that each successive state is obtained from the preceding state by means of an allowable action (operation applied to one or more expressions in the preceding state).

Part (d) really includes the first three parts, so it may be taken to be a sufficient definition of a problem solution. However, part (d) appears to place primary emphasis on the sequencing of actions, and in many problems it is the specification of givens or operations that constitutes the main source of difficulty in the problem. Thus, it is important to give these matters proper emphasis.

A simple and completely equivalent definition of a solution is to say that a solution is a sequence of allowable actions that produces a completely specified goal expression.

In Instant Insanity, a solution could be considered to consist of some given configuration of the four cubes, followed by a sequence of different configurations of the cubes, each of which was obtained by an allowable operation from the previous configuration, and ending with a configuration that satisfies the goal of having each of the four colors represented once on each of the four sides of the row of four cubes.

In a chess problem, a solution consists of some given board configuration, followed by a sequence of board configurations, each of which is derived from the previous configuration by an allowable move, and ending with a checkmate configuration. If the problem asserts that this solution is to be accomplished with some restrictions on the number of moves, then the description of the problem state must include a move counter that is increased by one on every move. The terminal expression must not only be a checkmate position, but the move counter must be less than or equal to some value. Chess problems are often optimization problems, in which the different solutions have different values depending upon how few moves they require.

In algebraic find problems or logical proof problems, the solution consists of a sequence of states such that (a) the given state is the conjunction of all the givens, (b) each successive state is derived from the previous state by adding an expression that has been obtained by applying an allowable operation to one or more of the previously obtained expressions, (c) the goal state includes a completely specified goal expression. When there are several given expressions, the most common practice is to write down the given expressions only as soon as they are needed for some operation. This procedure makes it easier for the reader to follow the proof, but I think it is more logical to regard all the givens as having been written down in the given problem state. If there is some psychological benefit in writing them down again in problems involving only nondestructive operations, of course you should do it. But I do not think this writing exercise should influence your definition of a problem solution.

STATE-ACTION TREE

Although the solution of a problem can be defined in terms of *either* a sequence of actions or a sequence of states (terminating with the achievement of the goal), it is very useful to represent *both* the possible sequences of actions and the possible sequences of states in a common diagram, which could be called a *state-action tree* for a problem. An example of such a tree is shown in Fig. 2-3.

In a state-action tree, the *nodes* or branch points of the tree represent all the *possibly* different problem states that could result from all the different action sequences. The concept of a node in a state-action tree differs from the concept of a problem state in a somewhat subtle, but important, way. To be sure, every node represents a state of the problem, but two distinct nodes do not necessarily represent two distinct or different states of the problem. That is, two or more action sequences, which result in two different nodes, may result in two identical

No. possible states

State level

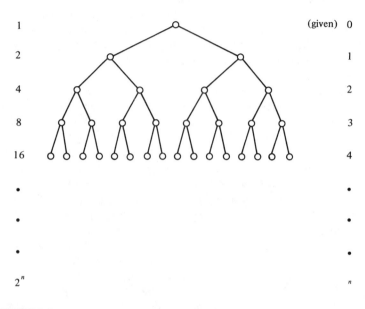

No. possible states	State level
1	(given) 0
2	1
4	2
8	3
16	4
•	•
•	•
•	•
2^n	n

FIGURE 2-3
State-action tree for a problem with two possible actions at each state, showing how the number of possible "terminal" states at level n (equaling the number of different action sequences that are n actions in length) increases geometrically with n.

problem states. Strictly speaking, a node represents the sequence of actions or the sequence of states that led up to it, not the problem state achieved by that sequence of actions or states. However, as long as you bear in mind that distinct nodes do not necessarily represent distinct problem states, there is no harm in considering a node to represent a state, rather than the sequence of actions or states that led up to it.

The *branches* from each node represent the different actions that could be selected at that node. Obviously, the actions possible at each node need not be similar to the actions possible at any other node, but in many problems the actions that are possible at each node fall into the same action classes or operations, with only the available operands being different; however, this similarity is not true of every problem. In addition, the number of possible actions at each node need not be equal either at the same level or across different levels.

These possible differences from node to node do not alter the primary lesson to be learned from examining a state-action tree—

namely, how rapidly the number of possible nodes or action sequences
increases in such a tree as a function of level, that is, the length of
the prior action sequence. If m actions occur at each node, then there
are m^n possible action (or state) sequences terminating at level n. Each
of these different action (state) sequences is represented by a node at
level n in the state-action tree, so there are m^n different nodes at level n.

This geometric (discrete exponential) increase is perhaps the single
most important fact to consider in developing problem-solving methods.
To solve a problem you must state the exact sequence of actions
(states) that results in the goal, and many problems require a moder-
ately long sequence of actions to accomplish the goal. Thus, we are
often faced with a search among an extremely large number of alternative action sequences. In these cases, we must "prune the tree"
so that there are not so many possible action sequences to investigate. But, of course, we must prune in such a manner that we do not
cut off all the branches that have "fruit," that is, states including
the goal.

If you had no basis for choosing between the alternative actions at
each node, if all the nodes at all levels represented distinct states
(distinct sets of expressions), and if only one of the states (up to and
including level n) included the goal, then there would be no way to
prune the tree and reduce the search. However, in most problems, it is
possible to prune the tree.

Different sequences of actions often result in equivalent problem
states, allowing you to combine nodes, prune branches, construct
equivalent reduced state-action trees, and so on (for example, classificatory trial and error and macroaction in Chapter 4). Usually, there
are good reasons for choosing certain actions at any node and ignoring other actions and the branches they generate (for example, state
evaluation and hill climbing in Chapter 5). Frequently, a large problem
can be broken up into subproblems, thereby transforming a large tree
into several smaller trees, with a great reduction in the total number of
branches (for example, subgoals in Chapter 6). Sometimes, a much
smaller tree results from trying to get from the goal back to the givens,
rather than the reverse (for example, working backward in Chapter 8).

Problems with multiple given states can be represented by as many
state-action trees as there are possible given states. In some problems,
the principal task is to choose among the given states (alternative sets
of givens), the one or more given states whose state-action trees contain a goal state. Often these problems require only a very short action
sequence to achieve the goal, once the correct given state has been
selected. In such problems, the main difficulty is to find the correct
type of tree in a large forest; climbing the tree may pose only a minor

problem. The method of contradiction discussed in Chapter 7 is often useful for these problems.

There is a special case of problems with multiple given states that occurs quite frequently and is of particular interest. In these problems, the solver has the option of considering state A to be given and state B to be the goal or of considering state B to be given and state A to be the goal. This kind of equivalence between two problems occurs where inverse operations exist for all operations. One problem of this type was discussed earlier in the chapter — namely, the equivalence of deriving $x = 3$ from $4x + 5 = 17$ or vice versa.

3

Inference

Virtually all problems present some of the relevant information in implicit, rather than explicit, form. That is, some of the information concerning givens, operations, or occasionally even goals is presented in a subtle manner that may not strongly attract your attention, unless you know what to look for. In a sense, this situation might be said to be poor communication of the components of a problem. Why do not the people who make up problems simply do a better job of communicating the relevant information?

I would agree that, in some cases, problems used for teaching purposes could be improved by making the relevant information very clear. In these cases the problem is difficult enough in explicit form without the added difficulty of the relevant information being presented implicitly. However, when you are posing and solving mathematical, scientific, and engineering problems for yourself in some real-life endeavor, your own initial posing of problems will contain implicit statements of information. Unless you know how to analyze a problem for implicit information, you will have difficulty solving actual problems later on.

Problems often evolve from (a) vaguely formulated to (b) semi-precisely formulated to (c) precisely but partly implicitly formulated to (d) precisely and explicitly formulated stages. It is very important for problem solvers to know what kinds of implicit information to look for in problems, because this information is often a critical step in problem solving, whether in school or in life. Furthermore, even when all the givens and operations are explicitly presented in the problem, it is, of course, necessary to transform the givens by means of the operations in some way in order to solve the problem. The solver must make inferences, draw conclusions, from the given information, a process that is, in essence, rendering explicit the statements that were (in a somewhat different sense) only implicit in the givens.

When *implicit information* refers to the consequences of given information, it is a somewhat different use of the term than when it refers to information not contained in the explicit statement of the problem (although, by convention, one is supposed to know that it is part of the information in the problem). However, there are all degrees of explicit mention of implicit information found in different problems. For example, a problem might refer to even numbers. In one sense, this statement is explicit mention of even numbers from which one can draw the inference that, if n is an integer and an even number, then it can be expressed as $2m$, where m is also an integer. However, the definition of even numbers is not presented explicitly in the problem and must be supplied from memory. This sort of semiexplicit, semiimplicit presentation of information occurs all the time in problems. Thus, it is probably not too useful to distinguish between the drawing of conclusions from different degrees of implicitly versus explicitly presented information.

Drawing inferences from implicitly or explicitly presented information is essentially random trial and error, unless some criteria are specified regarding which inferences (more generally, which transformations of the goal or the given information) should be made first. There are essentially two criteria that can be formulated semiprecisely, but not completely precisely, at the present time. The first criterion is that the inferences should be those that you have frequently made in the past from the same type of information. You assume that the properties that proved useful in the past will most likely prove useful in the present problem. The second criterion is that the inferences you draw should be those inferences that are concerned with properties mentioned in the goal, the givens, or in previously derived consequences of the goal or the givens. Inferences that satisfy this second criterion are likely to combine with other information to yield still further inferences.

Thus, the general problem-solving method described in this chapter may be stated as follows: *Draw inferences from explicitly and implicitly presented information that satisfy one or both of the following two criteria: (a) the inferences have frequently been made in the past from the same type of information; (b) the inferences are concerned with properties (variables, terms, expressions, and so on) that appear in the goal, the givens, or inferences from the goal and the givens.* Throughout the rest of the book, the expression "drawing inferences" will be used to refer to the above statement of the method — namely, drawing inferences that satisfy one or both of the previously stated criteria.

Drawing inferences (more generally, making transformations of the goal or the givens) is probably the first problem-solving method you should employ in attempting to solve a problem. You are essentially expanding the goal or the givens by bringing to bear all of the knowledge you have concerning this problem in your memory. Frequently, problems are quite simply solved, once all the relevant information is retrieved from memory, in the drawing of inferences from explicitly and implicitly presented information. Most people do make frequent use of the inference method, at least in connection with drawing inferences from givens. (This procedure is often thought to be random trial and error, but this characterization is largely inaccurate, since people's inferences usually do meet one or both of the stated criteria.) The general problem-solving methods discussed later in the book are somewhat less universally used by human problem solvers, but the discussion of them should not lead you to ignore the basic inference method. For this reason, this method is the first general problem-solving method discussed in this book. Furthermore, a greater understanding of how the inference method operates and an awareness of some illustrative use can greatly facilitate your proficiency in using the method, particularly with respect to inferences from the goal information, which people do not pay enough attention to. People have a bias to start at the beginning, which they take to mean the givens. This bias is often inappropriate in problem solving, since the goal is frequently a better beginning point than the givens.

So-called *insight problems* are often problems in which the principal step in solution is to draw the appropriate inference from certain explicitly or implicitly presented information. Very few steps are required to solve the problem. What is necessary is to make that one critical transformation of the givens that essentially solves the problem. Difficult insight problems are often difficult precisely because they require you to draw an inference that is not too close to the top of your hierarchy of inferences from this type of given information [criterion (a)]. Obviously, the more you have stored in your memory concerning the principal inferences to be drawn from the types of given information contained in the problem, the more likely you are to be able to achieve the critical insight. However, whatever your level of specific knowledge concerning the given information, greater understanding and experience in the use of the inference method will increase your chances of systematically discovering the required insight in the course of drawing inferences concerning properties of the given information. Just knowing that what you are doing is surely not random trial and error may cause you to go further and further down the list of inferences to be made from the information in the problem, rather than giving up this approach after the first few inferences fail. With the knowledge of problem-solving methods contained in this book and experience in applying them to the solution of problems, you can gradually develop a fairly accurate intuition as to which problems are insight problems and thus most suited to the inference method and not to other problem-solving methods. If you classify a problem as an insight problem, then you should continue drawing inferences (rather than use other methods) for a longer period of time than if you do not classify it as an insight problem.

Of course, drawing inferences (including explicit representation of implicit information) is often an important part of solving any problem, not just insight problems. Insight problems are simply those in which inference is the principal or only method employed in solving them. In noninsight problems, you should stop using the inference method when you "run out of gas" using the method—that is, when you find it difficult to draw from the given information any new conclusions that seem to have any likelihood of being useful in solving the problem. In noninsight problems, you should then go on to consider employing other general problem-solving methods, using the expanded set of given information provided by the inference method. In insight problems, when you run out of gas, you should go back and try over and over again to look at the problem from a different point of view to yield additional new inferences.

The discussion of inference and implicit information naturally divides into three sections. First, givens may be, to some extent, stated implicitly and, in any event, can usually be expanded considerably by use of the inference method. Second, operations are not always explicitly stated. Third, the goal of the problem is occasionally not completely clear, and the solver must get a precise and correct definition of the goal. In addition, it is often helpful to specify the properties of the goal in more detail. This procedure frequently involves drawing inferences from presented information (givens and goal), including explicit symbolic or diagrammatic representation of information that may appear only implicitly in the problem.

GIVENS

The problems at the end of a section in a textbook are there to test the reader's knowledge of the material presented in that section. Each problem, then, includes all of the given assumptions, proved theorems, and operations that appeared in the section as well as the particular givens of the particular problem. In addition, some previous material presented in the book may be relevant to solving the problem, and certain background knowledge from other books may also be needed. Such background information concerning givens and operations is one kind of implicit information in problems.

You should be aware of this kind of implicit information in problems, and take care to master background subject matter before proceeding on to courses that have this background as a prerequisite. If you have not fully understood what was presented previously in the course or what was presented in relevant background courses, you should face this fact and go back to learn the relevant prior material, either simultaneously with or instead of taking a subsequent course. It is lunacy to go on to more advanced courses without a reasonably clear understanding of the relevant background material. The general problem-solving methods taught in this book will not substitute for lack of the relevant knowledge.

It is true that you can understand the relevant material and not be able to solve problems for lack of understanding of general problem-solving methods. However, you will also fail to solve problems if you lack the relevant knowledge, no matter how skillful a problem solver you are. In today's schools a C or even a B in a course may represent an inadequate level of understanding for going on to more advanced courses, and the conscientious student should recognize this fact and act accordingly.

In addition to background information, there is another kind of implicit problem information that the skilled problem solver can come to recognize rather easily, sometimes greatly facilitating solution. This other kind of implicit information concerns the properties possessed by each of the givens or operations in a problem. When a familiar object or activity is presented in a problem, all of the known properties of that object or activity (including all its known relations to other objects or activities) are usually considered to be part of the given information. There may be no question that everyone who works on the problem knows all of the relevant properties of all the givens and operations in the problem. That is, no specialized background knowledge is required. However, amateur problem solvers frequently fail to ask themselves what they know about the givens and operations in a problem from their own past experience. Insight problems are very often problems that require one to notice — which means represent explicitly — properties of givens presented in the problem.

Of course, many of the implicit properties of the givens are irrelevant to solving the problem. We know that most people have two legs, two arms, two eyes, skin, hair, a nose, a mouth, and so on, but most of these properties are irrelevant to the solution of any single problem where people are included in the given information. Such irrelevant properties should be ignored, and problem solvers are usually able to reject such truly irrelevant implicit properties. The difficulty usually comes in abstracting or consciously considering the possibly relevant implicit properties. Some examples are described in the following subsections.

Numerical Properties

Whenever numbers are involved in a problem in any way, you should consider whether the known properties of the kind of numbers involved in the problem might be of any value in solving the problem. For example, if some number is known to be a positive integer, then it cannot be negative, zero, or a fraction. If an integer, n, is known to be even, then it can be expressed as $n = 2m$, where m is also an integer, or as $n = 2^s p$, where s is an integer and p is an odd integer. If an integer, n, is known to be odd, then it can be expressed as $n = 2m + 1$, where m is an integer, or $n = 2^s p + 1$, where s is an integer and p is an odd integer.

A somewhat famous example in the psychology of problem solving of the abstraction of numerical properties comes in the *13 problem* of Karl Duncker (1945, p. 31). The problem can be stated as follows:

> Prove that all six-place numbers of the form *abcabc* (for example, 416416 or 258258) are divisible (evenly) by 13.

Stop reading and try to solve this problem, then read on.

You might try a variety of special cases, verifying that in every case the number was divisible by 13, but that would probably not suggest how to prove the theorem in general. The critical step is to inquire whether you know any numerical properties of a number of the form *abcabc*. If you could not solve this problem before, stop reading and try again by abstracting numerical properties of numbers of the form *abcabc*.

If you still could not solve the problem, consider whether you could factor a number of the form *abcabc* into a product of other numbers. Now stop reading and try again.

In factoring the number, you no doubt determined that *abcabc* = (*abc*)(1001), for all numbers of the form *abc* and therefore for all numbers of the form *abcabc*. Now, of course, 1001 is divisible (evenly) by 13, so (*abc*)(1001) is divisible by 13, and the theorem is proved.

Furthermore, the factoring of *abcabc* into *abc*(1001) can be achieved quite automatically by representing the numerical properties of *abcabc* in the following standard way (for which *abcabc* is really the conventional abbreviation):

$$abcabc = (a \cdot 10^5) + (b \cdot 10^4) + (c \cdot 10^3) + (a \cdot 10^2) + (b \cdot 10) + (c)$$

$$= a \cdot (10^5 + 10^2) + b(10^4 + 10) + c(10^3 + 1)$$

$$= a \cdot 10^2 (10^3 + 1) + b \cdot 10(10^3 + 1) + c(10^3 + 1)$$

$$= (1001)(a \cdot 10^2 + b \cdot 10 + c) = (1001)(abc)$$

Topological Properties

Topology is concerned with the properties of geometric figures that remain unaltered when the figures are stretched, shrunk, and twisted in any regular or irregular way. For example, consider the square shown in Fig. 3-1. Imagine that the square was drawn on a sheet of very flexible rubber and that it was stretched so that the square looked like that shown at right in the figure. What properties remain invariant under the stretching, shrinking, and twisting of the rubber sheet? Actually, a number of properties are unchanged. Points inside the figure remain inside, points outside the figure remain outside, and points on the edges (lines) of the figure remain on the edges. If you consider that the figure has only four points — namely, the four vertices *A*, *B*, *C*, and *D* — and that the edges are defined merely as unordered pairs of the vertex points, then the set of points and the set of edges (unordered pairs of points) has not been changed by the distortion either.

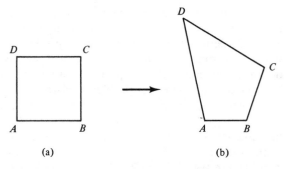

FIGURE 3-1
Distorting a square drawn on a rubber sheet to illustrate
the topological properties of a figure (those properties
that are unchanged by stretching, shrinking, and twisting).

Consider a figure with several faces or regions entirely enclosed
by lines with no interior lines, such as the three-face figure shown in
Fig. 3-2. All of the invariants described in the preceding paragraph for
a single-face figure obtain for the multiface figure. In addition, the
faces that border on each other (have a common edge) still border on
exactly the same faces after the distortion. Thus, if you constructed
the set of unordered pairs of faces that border on each other—namely,
(f, g) and (g, h)—this set would remain invariant under stretching,
shrinking, and twisting.

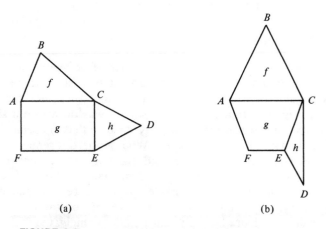

FIGURE 3-2
Distorting a three-face figure drawn on a rubber sheet to
illustrate topological properties. Faces are represented by
f, g, and h. Vertices are represented by A, B, C, D, E, and F.

One of my favorite problems involves the property of the bordering (direct connection) of faces in an important way. This is the *notched-checkerboard problem*:

> You are given a checkerboard and 32 dominoes. Each domino covers exactly two adjacent squares on the board. Thus, the 32 dominoes can cover all 64 squares of the checkerboard. Now suppose two squares are cut off at diagonally opposite corners of the board (see Fig. 3.3). Is it possible to place 31 dominoes on the board so that all of the 62 remaining squares are covered? If so, show how it can be done. If not, prove it impossible.

Stop reading and try to solve this problem.

If you could not solve it, consider the following hint. This problem primarily involves use of the inference method to explicitly represent certain properties of the checkerboard and dominoes that are only

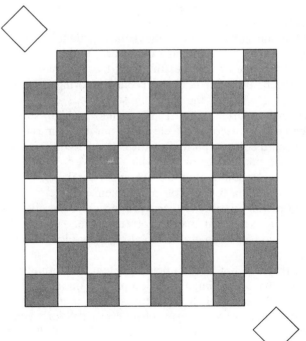

FIGURE 3-3
The notched checkerboard.

implicitly presented in the present problem. Once the appropriate property or properties are recognized, the solution to the problem is obvious. Now stop reading and try to solve the problem, if you could not do so before.

The critical property is that of the two squares of the checkerboard that are covered by any domino. What are some of the properties of any such two squares? If you have not yet solved the problem, stop reading and try again, considering this hint.

The critical properties of the two squares covered by any domino can be expressed in terms of the colors of these two squares. What are the colors of the two squares covered by any domino on a checkerboard? If you have not yet solved the problem, stop reading and try again, considering this hint.

The key insight required to solve the notched-checkerboard problem is to notice that a domino covers two squares that are *always* of different colors (that is, one black and one white). Since the diagonally opposite corner squares are of the same color, there are now 30 squares of one color and 32 squares of the other color, and obviously the 62 squares cannot be covered by 31 dominoes.

What has intrigued me most about the problem is this: the impossibility of covering the remaining 62 squares with 31 dominoes can be proved irrespective of whether the eight-by-eight matrix is presented as a checkerboard with a checkerboard coloring pattern and even irrespective of whether the problem solver has ever experienced a checkerboard coloring pattern. But what problem-solving methcd would lead one to discover the elegant proof that comes from imposing a checkerboard coloring pattern on the matrix? Is this kind of ingenious idea a chance happening, or something only very brilliant people can think of, using methods that are not understandable by others? I do not think so. I think that use of the problem-solving method of representing all of the possibly relevant properties of the givens in a problem makes it likely that many problem solvers would discover the elegant solution of even the notched eight-by-eight colorless matrix problem.

I think that it is not likely a person unfamiliar with a checkerboard coloring pattern would impose such a pattern on a colorless eight-by-eight matrix. However, I think that it is likely that a person would do something equivalent to imposing checkerboard coloring on the matrix, as follows. Using the method of trying to represent all of the possibly relevant properties of the givens in the problem, one would eventually label the squares in the eight-by-eight matrix in ordered-pair (coordinate) notation, as shown in Fig. 3-4. Now one might eventually

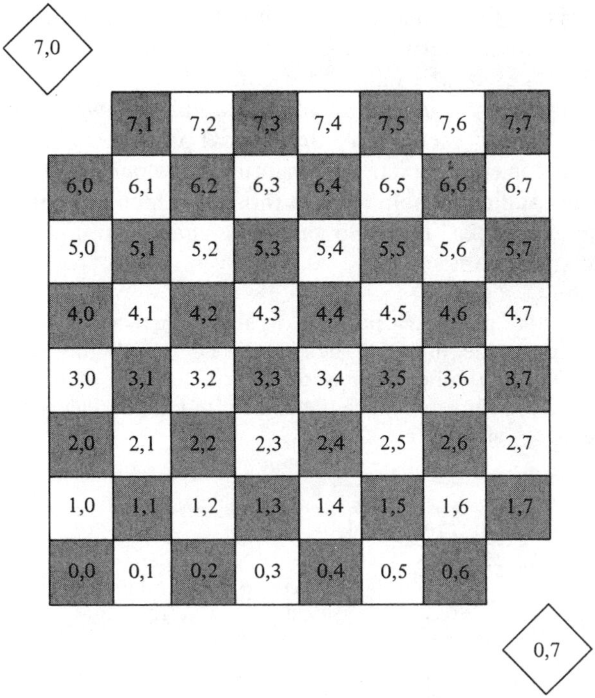

7,0

7,1	7,2	7,3	7,4	7,5	7,6	7,7
6,0	6,1	6,2	6,3	6,4	6,5	6,6	6,7
5,0	5,1	5,2	5,3	5,4	5,5	5,6	5,7
4,0	4,1	4,2	4,3	4,4	4,5	4,6	4,7
3,0	3,1	3,2	3,3	3,4	3,5	3,6	3,7
2,0	2,1	2,2	2,3	2,4	2,5	2,6	2,7
1,0	1,1	1,2	1,3	1,4	1,5	1,6	1,7
0,0	0,1	0,2	0,3	0,4	0,5	0,6

0,7

FIGURE 3-4
The notched checkerboard with ordered-pair (coordinate)
labeling of the squares.

look for some property common to all pairs of squares that a single domino could cover. If the idea occurred to one to look for this kind of property, then having labeled the squares in ordered-pair (coordinate) notation, it is likely that one would see that a domino must cover two squares, one of whose coordinate sums is odd and the other even. Since the diagonally opposite squares of the matrix both have either an odd or an even coordinate sum, the notched matrix cannot be covered by the 31 dominoes. The solution is in every way equivalent to that given for the notched checkerboard using the color property but in no way requires one to invent some special labeling scheme such as a checkerboard coloring pattern. Only the very generally useful and familiar coordinate labeling scheme is needed.

Let us examine why this problem is an example of the abstraction of the topological properties of a figure. A domino covers two faces that border on each other in a complex figure composed of faces with a very special type of bordering structure. It is the bordering structure of the matrix of faces that is represented by the coordinate labeling scheme (or the checkerboard coloring pattern), and the shapes and sizes of the faces or the matrix are completely irrelevant. Thus, the notched eight-by-eight matrix problem is a problem where the critical properties to be represented are topological properties.

Other problems in which representing topological information is important for achieving solution are those in which a block is cut into component subblocks. The following *cube-cutting problem* is, I guess, the classic such problem:

> You are working with a power saw and wish to cut a wooden cube, 3 inches on a side, into 27 1-inch cubes. You can do this by making six cuts through the cube, keeping the pieces together in the cube shape (see Fig. 3-5). Can you reduce the number of necessary cuts by rearranging the pieces after each cut?

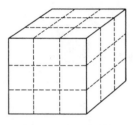

FIGURE 3-5
Slicing a 3-by-3-by-3-inch cube into 27 subcubes.

Stop reading and try to solve the problem.

Consider the 3-by-3-by-3-inch cube to be already divided into its 27 component cubes but still stacked in such a way as to form a 3-by-3-by-3 cube. The important topological properties of such a structure are concerned with the vertices, the edges, and the faces of the component cubes. If you did not solve the problem, stop reading and try again.

Among the important topological properties the one most likely to be relevant to the solution of the present problem concerns the faces of the component cubes, since the power saw essentially separates the faces of certain component cubes from the faces of other component cubes. If you have so far not solved the problem, stop reading and try again, using this hint.

The 27 component cubes fall into several classes on the basis of how many of their faces (sides) border on other component cubes versus how many are parts of the exterior faces of the 3-by-3-by-3 cube. Classify the component cubes by this criterion and consider this information in relation to the solution of the problem. If you have not solved the problem thus far, stop reading and try again.

There are four classes of cubes with respect to the property of the number of "interior" faces (that is, the number of faces that border on faces of other component cubes and thus must be cut). There are the 8 corner cubes that have only three interior faces; there are the 12 edge cubes that have four interior faces; there are the 6 face cubes that have five interior faces; and there is the one center cube that has six interior faces (and is totally hidden from view in the 3-by-3-by-3 cube). A cross-sectional diagram of the cube representing the number of interior faces for each component cube is shown in Fig. 3-6. If you have not solved the problem thus far, consider the information in Fig. 3-6 and try again.

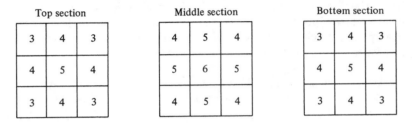

Top section		
3	4	3
4	5	4
3	4	3

Middle section		
4	5	4
5	6	5
4	5	4

Bottom section		
3	4	3
4	5	4
3	4	3

FIGURE 3-6
The number of interior faces (needing to be cut) for each component cube of the 3-by-3-by-3-inch cube.

The key insight required to solve block-cutting problems in general and this cube-cutting problem in particular is to focus on that subblock which has the greatest number of faces that must be cut in order to separate it from the other subblocks. The reason for focusing on the subblock with the largest number of faces to cut is that the number of such faces on this block sets a minimum to the number of cuts that must be made. It is obvious why this is so, since under no circumstances can one cut more than one face of a subblock at a time. In the case of the cube, this fact means focusing on the most central cube, which has no exposed faces to begin with. This cube has six faces that must be cut, and therefore no fewer than six cuts will solve the problem. Since we know by inspection that six cuts will solve the problem, the number of cuts that is required is exactly six.

The same principle can be applied to a large class of other problems to set a minimum on the number of cuts that are required. For example, a cube cut into four subcubes, as in Fig. 3-7, requires three cuts, since each of the four subcubes has three unexposed faces that must be cut.

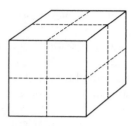

FIGURE 3-7
Slicing a 2-by-2-by-2-inch cube
into subcubes.

Operations

Many practical problems require you to think of a type of operation
that will solve the problem. The operation is usually one with which
you would be quite familiar, but thinking of that operation may be far
from trivial. Nevertheless, it is probably of some help to be explicitly
aware of the possibility of implicit operations and to have some
examples of such problems in your mind. One example is the well-
known *radiation problem* of Duncker (1945, p. 1):

> Given a human being with an inoperable stomach tumor, and rays which
> destroy organic tissue at sufficient intensity, by what procedure can one
> free him of the tumor by these rays and at the same time avoid destroying
> the healthy tissue which surrounds it?

There are a number of plausible solutions, each of which involves
thinking of some operation not specified in the statement of the problem. For example, the rays might be focused from several sources so
that they intersected in the region of the tumor. A single source of
radiation could be rotated around the body so that all the beams intersected in the region of the tumor. Perhaps a source of radiation could
be implanted inside the tumor.

In some formal problems with a precisely delimited set of operations,
the properties of one or more of the operations may be somewhat implicit. Failure to achieve a completely explicit and accurate understanding of the properties of such operations may block solution of
the problem. As an example of this type of situation, consider the
following *one-heavy-coin problem*:

> You have a pile of 24 coins. Twenty-three of these coins have the same
> weight, and one is heavier than the others. Your task is to determine
> which coin is heavier and to do so in the minimum number of weighings.
> You are given a beam balance (scale), which will compare the weights
> of any two sets of coins out of the total set of 24 coins.

Stop reading and try to solve the problem. Consider the properties of the weighing operation if a beam balance is used. What kind of information does the beam balance provide concerning relative weights in any two sets of coins? How many different outcomes are there to a weighing on a beam balance? If you have not solved the problem thus far, stop reading and try again.

A beam balance actually has three different outcomes, not two— namely, the left pan is heavier, lighter, or equal in weight to the right pan. Since there are three different outcomes to weighing on a beam balance, it is at least theoretically possible that a beam balance could provide one with an answer as to which of three subsets of coins contains the heavy coin (not just deciding which of two subsets contains the heavy coin). Consider this hint, and you should easily be able to solve the problem.

A beam balance has two pans and compares the weights of two sets of coins. For this reason, many people assume that the operation they have available to solve the problem is essentially to ask which of two (equally large) sets of coins contains the heavy coin. Accordingly, they reason that the optimal strategy must be to divide the total set of coins in half and weigh one half against the other half (12 coins against 12 coins). Then, having determined which set of 12 coins contains the heavy coin, they proceed to divide that set in half and weigh 6 coins against 6 coins, then 3 against 3, and finally 1 against 1, or 2 against 2 followed by 1 against 1.

When the number of coins remaining is only three, it might occur to a person that one's original characterization of the operation as a two-way question was in error. However, sometimes even this simple terminal problem does not necessarily indicate to the problem solver that the beam balance can actually provide one with an answer to a three-way question, if the coins are divided into three piles (two of which are equal) on each and every weighing. This procedure is, of course, the solution to the problem—namely, you should first weigh two sets of eight coins against each other. If one pan is heavier than the other, then it contains the heavy coin. If the two pans balance, then the heavy coin is in the remaining set of eight coins that was left off the pans of the balance scale. In any case, you find out which subset of eight coins contains the heavy coin. You then continue to partition the remaining set of eight coins into three parts by weighing three coins against three coins. No matter which subset of coins contains the heavy coin, the answer will be found in one additional weighing, or three weighings in all. By contrast, dividing the set into two equal parts (whenever possible) requires four weighings.

If you are careful to explicitly state the properties of the operations available in a problem, then you will be more likely to avoid the inaccurate characterizations that frequently occur in problems such as this one.

GOALS

Occasionally in school and more frequently in real-life problems, the goal of a problem is not completely clear. Obviously, an important step in representing the information in a problem is to be sure you have a precise and correct definition of the goal. It is often worthwhile to question whether or not you understood the goal correctly, since it is sometimes easy to make a mistake in this regard. As an example, consider the following logic problem:

> The country of Marr is inhabited by two types of people, liars and truars (truth tellers). Liars always lie and truars always tell the truth. As the newly appointed United States ambassador to Marr, you have been invited to a local cocktail party. While consuming some of the native spirits, you are engaged in conversation with three of Marr's most prominent citizens: Joan Landill, Shawn Farrar, and Peter Gant. At one point in the conversation Joan remarks that Shawn and Peter are both liars. Shawn vehemently denies that he is a liar, but Peter replies that Shawn is indeed a liar. From this information can you determine how many of the three are liars and how many are truars?

Stop reading and try to solve this problem, then read on.

In solving logic problems of this type, it is a common procedure to list all of the possibilities in the form of one or more tables. For example, in this problem one might attempt to fill out the following table:

Person	Liar	Truar
Joan	____	____
Shawn	____	____
Peter	____	____

In a problem very similar to this one, a student in one of my problem-solving classes attempted to solve the problem by filling out just such a table. Perhaps you did this in the present problem. However, in translating the problem into this form, a subtle transformation has taken place with respect to the goal—namely, the goal has been changed from determining *how many* of the three are liars to determining *whether each* of the three persons is a liar or a truar. If you try to answer the new version of the problem, you will never be able to reach a solution! All you must determine is how many of the three are liars. The correct answer is a number: 0, 1, 2, or 3. The names of the people are largely irrelevant information, added to make the problem appear more interesting and simultaneously to act as a distraction. If you have not yet solved the problem, stop reading and try again, then read on.

In fact, it is impossible to determine whether Shawn is a liar and it is also impossible to determine whether Peter is a liar. However, one can conclude that either Shawn is a liar and Peter a truar or Shawn is a truar and Peter a liar. Either way one and only one of the men is a liar. Thus, Joan must be a liar. This conclusion implies that there are exactly two liars and one truar in the group of three natives to whom you are talking. This answer solves the original problem but does not allow you to completely fill out the table.

In addition to reading and rereading a problem to avoid misunderstanding the goal, you should have a clear, precise statement of the goal, rather than some vague formulation of it. Sometimes vague statements of the goal are partly or completely due to unclear statements of the goal in the problem, and sometimes the vague formulation is due partly or completely to sloppy reformulation by the problem solver. In either case, a vague formulation of the goal may do considerable harm when you attempt to solve a problem. For example, consider the *cheap-necklace problem*:

> You are given four separate pieces of chain that are each three links in length. It costs 2¢ to open a link and 3¢ to close a link. All links are closed at the beginning of the problem. Your goal is to obtain a single closed chain, using all links, at a cost of no more than 15¢.

The goal is to obtain a single closed chain, using all links. But what does that mean? Is a single closed chain a simple loop or circle? Or would multiple loops be satisfactory? Is it conceivable that a closed chain might mean a long chain formed without joining the ends together in a loop? Could there be some other variations on these possibilities? Until you can decide which of the reasonable possibilities constitutes the actual goal of the problem, you do not really know what the problem is and cannot expect to make much progress in solving it. Some people object to such deliberately vague statements of goals, but vagueness regarding goals is often a feature of real problem solving and you probably should get some experience in dealing with this kind of problem. Whether in school or elsewhere, the lesson is that you should be sure you have a precisely formulated and accurate understanding of the goal.

In addition to having a precise and accurate understanding of the goal, you will frequently find it helpful to have a more *detailed* representation and understanding of the goal than may be provided in the original statement of the problem. As Polya (1962, p. 7) has emphasized, it may be useful to imagine for a moment that you have already solved the problem and ask yourself, "What would I have?" Polya variously calls this exercise wishful thinking or taking the problem as solved. Whatever one calls this type of thinking, it involves some sort of increase in the explicit representation of the goal either in symbolic (verbal) or diagrammatic form. The usefulness of increasing the specification of the goal may involve little more than introducing names (labels, symbols) for concepts that appear in the goal but are not explicitly represented in that form in the original statement of the problem. Thus, one purpose of increasing the specification of the goal is to introduce the necessary working concepts for reaching the goal. Another purpose is to derive some additional properties possessed by the goal, either by rigorous inference from the information contained in the goal and/or the givens of the problem or by representing a reasonable conjecture (guess) based on one or another heuristic consideration. In either case, deriving additional properties of the goal may make it easier to reach the goal because then you have a more specific idea of the different components that you are attempting to achieve.

As an example of a problem where it is useful to represent the goal explicitly in a diagrammatic form, consider the following geometry construction problem:

Given an acute angle *UVW* and a point *P* within the angle, use a compass and straightedge to construct a segment *QR* passing through *P*, such that *QP* and *PR* stand in the ratio 2:1, *Q* and *R* lying on *UV* and *VW*, respectively.

In solving this problem, it is, of course, useful to represent the acute *UVW* and the point *P* within the angle as shown in Fig. 3-8. However, although we do not yet know exactly where point *Q* lies on the line *UV* or where point *R* lies on the line *VW*, it is useful to explicitly represent the goal line *QR*. This representation is done by drawing in a hypothetical dashed line, as shown in Fig. 3-8. The solution to this problem will be discussed in more detail in Chapter 4, but the advantage of explicit representation of the line *QR* is that you are more

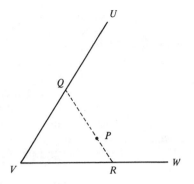

FIGURE 3-8
Explicit representation of goal line.

likely to see how to construct similar triangles involving the line segment *QR*, and the construction of these similar triangles is a critical step in solving the problem. Although the placement of the line *QR* in Fig. 3-8 is certainly not expected to be exactly correct, it gives you a more explicit representation of what the final goal would look like. In this case, that makes it much more probable that you will see certain relations that are critical in solving the problem.

Increasing the specificity of the goal generally means more than merely drawing an extra line or two in a figure or introducing a few new symbols (important though this purely representative aspect may be). Often it means deriving additional properties possessed by the goal, using either the statement of properties of the goal as given in the original problem or possibly also using given information to derive properties of the goal (without necessarily achieving the entire goal). A marvelous example of the importance of deriving properties of the goal is provided by the following plane-geometry problem:

> Can two triangles have five of their six parts (three sides and three angles) be equal and yet the triangles not be congruent?

Stop reading and try to solve this problem, using the problem-solving method of explicit representation of the goal and deriving properties of the goal.

The first inference you might make is that, for two triangles to have five of their six parts equal, this goal subdivides into two alternative, more specifically stated goals — namely, the two triangles having three sides and two angles be equal or the two triangles having two sides and three angles be equal. Having explicitly represented both possibilities, it is easy to see that the first is impossible — namely, two triangles with three equal sides must be congruent, by a theorem of plane geometry. Thus, we need only consider the case where the two triangles have two sides equal and all three angles equal. If you did not solve the problem previously, stop reading and try again.

Another property that can be derived regarding the goal pair of triangles is that the two triangles must be similar. This property is a trivial restatement of the property already derived that the three angles are equal. Nevertheless, restating this property using the words "similar triangles" is quite helpful in bringing to mind a useful representation and the proper theorems regarding the relationship of corresponding parts in similar triangles. If you did not solve the problem so far, stop reading and try again.

If you had not already done so, you should have introduced some kind of diagrammatic representation of the two similar triangles that constitute the goal, such as the diagram illustrated in Fig. 3-9. Besides drawing the two similar triangles, you should also have labeled the sides in a manner that easily reflects which sides are corresponding, as is also shown in Fig. 3-9—namely, by using the same letter for corresponding sides and distinguishing the two triangles by the presence or absence of a prime. It is not strictly necessary for the solution of this problem to label the angles, but it does not hurt. If you have not solved the problem already, stop reading and try again.

Another inference that can be drawn regarding the goal is that the two equal sides in the triangles ABC and $A'B'C'$ will be noncorresponding sides. This conclusion is clearly true (from the method of contradiction, to be explained in Chapter 7), since, if the two sides were corresponding, we should have three equal angles and two equal corresponding sides in the two triangles, and such triangles are clearly congruent, by several theorems of plane geometry. If you still have not solved the problem, stop reading and try again.

Yet another relevant inference concerning the properties of the goal is that the ratios of all corresponding sides of similar triangles are

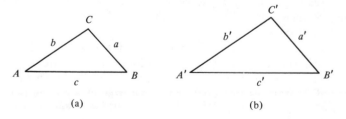

FIGURE 3-9
Two similar triangles with two equal sides (not corresponding sides).

equal. Thus, $a'/a = b'/b = c'/c$. If you have not solved the problem, stop reading and try again.

Another important set of inferences to be drawn from the goal are the inequalities that hold between the lengths of the different sides within each of the two triangles. Of course, it is completely arbitrary which side we decide is longest, next longest, and shortest. Nevertheless it is important to represent this information explicitly in solving this problem. As indicated in Fig. 3-9, we have assumed that $c \geq b \geq a$ and $c' \geq b' \geq a'$. If you have not solved the problem thus far, stop reading and try again.

Now it is useful to consider whether the goal triangles ABC and $A'B'C'$ could be equilateral or isosceles triangles. Since the triangles are similar, if one is equilateral the other is equilateral, and if one is isosceles the other is isosceles. Clearly, the triangles cannot be equilateral, since then all three sides of triangle ABC would have to be equal to all three sides of $A'B'C'$, or else all three sides of triangle ABC would have to be unequal to all three sides of triangle $A'B'C'$. Neither case satisfies the goal constraint of having two equal sides. In a somewhat similar manner, we can contradict the possibility that the triangles ABC and $A'B'C'$ are isosceles. If you have not already done so, stop reading and prove this and attempt to solve the rest of the problem.

In proving that the two triangles cannot be isosceles and in further work in connection with this problem, it is useful to derive another property of the goal triangles, namely, that triangle $A'B'C'$ is bigger than triangle ABC. Clearly, one is free to make this assumption without any loss of generality, since the labeling of the triangles is purely arbitrary. We can simply adopt a convention for convenience that the $A'B'C'$ triangle refers to the larger of the two similar triangles, no matter what pair of similar triangles we choose to work with. Incidentally, this trick of observing when one is free to assume certain relations without any loss of generality comes up often enough in problem solving to be worth taking special note of. In the present instance, if the triangles are isosceles, there are two possible cases: the two longest sides are equal ($c = b$ and $c' = b'$) or the two shortest sides are equal ($b = a$ and $b' = a'$). In the former case, the two larger sides (c' and b') of triangle $A'B'C'$ will be larger than any of the three sides of triangle ABC. Thus, there cannot be two sides of $A'B'C'$ equal to two sides of ABC. Similarly, in the latter case, there will be two sides (a and b) of triangle ABC that will be smaller than any of the three sides of triangle $A'B'C'$. So, in the latter case, there also cannot be two sides of triangle ABC that are equal to two sides of triangle

$A'B'C'$. Thus, we can assume that $c > b > a$ and $c' > b' > a'$. That is, the sides must have a strict inequality relationship among them, within any given triangle. Now, if you have not solved the problem already, stop reading and try again.

Again continuing to focus on the properties of the goal, we can derive which of the two sides of triangle ABC must be equal to which of the two sides of triangle $A'B'C'$. If you have not solved the problem, stop reading and answer the question concerning which sides of triangle ABC must be equal to which sides of triangle $A'B'C'$. Having answered that question, try again to solve the problem, if you have not done so already.

Since c' is the largest side of triangle $A'B'C'$ and triangle $A'B'C'$ is larger than triangle ABC ($c' > c$, $b' > b$, $a' > a$), in the goal triangles, b' must be equal to c and a' must be equal to b. That is, the largest side (c') of triangle $A'B'C'$ can have no side equal to it in triangle ABC, and the smallest side (a) of triangle ABC can have no side equal to it in triangle $A'B'C'$. Again, if you have not solved the problem already, stop reading and try again.

It is now helpful to represent another concept connected with the goal triangles, namely, the *expansion ratio* of the corresponding sides of the two triangles: $x = a'/a = b'/b = c'/c$. Using the relationships expressed in this series of equations, we can derive the equations $b' = xb$ and $a' = xa$. Recalling that in the goal triangles, b' must equal c and a' must equal b, we can derive the expressions $x = c/b$ and $x = b/a$. From this fact we conclude that in the goal triangles $c/b = b/a$. That is, the ratio of the large side to the middle side in the triangle ABC must be equal to the ratio of the middle side to the small side of triangle ABC.

Now all that is necessary is to realize that we have so completely specified the goal pair of triangles in this case that we have everything we need to solve the original problem. We know that two triangles can have five of their parts equal, provided that those parts are three angles and two sides and that the ratio of the large side of one triangle to the middle side is equal to the ratio of the middle side to the small side of the same triangle. Then clearly this relationship will hold for both triangles, since the triangles are similar. In addition to triangle ABC satisfying the relation $c/b = b/a$, it is also necessary that triangle ABC satisfy the triangle inequality, namely, $c < b + a$. However, there are an infinity of sets of three lines (a, b, c) that do indeed form a triangle (satisfy the triangle inequality) and are in the relation of $c/b = b/a$. For each such triangle (ABC), one can construct exactly one larger triangle (and one smaller triangle) such that five of the six parts of the two triangles are equal. The required expansion of the triangle ABC is obviously the factor $x = c/b = b/a$.

Thus, the original problem is solved. Note that in essence this was a construction problem, though it was phrased more as if it were an existence problem. However, in order to determine the existence of pairs of triangles having five of their six parts equal, it was necessary in this case to sketch a specific means by which such a pair of triangles could be constructed. The strategy used in this proof was to indicate a plan for constructing a pair of such triangles.

This problem provides a truly remarkable example of the importance in some problems of focusing on the goal and deriving properties of the goal (drawing inferences concerning the goal). The number of times that properties of the goal were represented or inferences were made concerning the goal in this problem was unusually large. Practically the entire problem-solving process consisted of representing or deriving properties of the goal, in this case. The reason for this extensive focusing on the goal was primarily that the givens were so unspecific — they were all the axioms and theorems of plane geometry. The only unique aspect of the problem that indicated what to select from all of our knowledge of plane geometry was the goal. Thus, we necessarily had to focus entirely on the goal, since it was the only unique aspect of the problem. Said another way, the goal provided us a unique beginning point from which to draw inferences, whereas if we had started from the givens our first step would have been to write down any axiom or theorem of plane geometry. Starting with the givens we would have had little idea where to proceed. Therefore, good strategy in this problem was to focus on the goal and make that goal progressively more and more specific to indicate exactly what aspects of plane geometry were relevant to solve the problem. In this case, once all the properties of the goal were explicitly represented, it was trivial to solve the problem from the beginning, by specifying a plan of construction.

A problem in a completely different context that illustrates the usefulness of focusing on the goal and deriving some of its properties at an early stage in problem solving is the following *63-link-chain* problem:

> Wanda the witch agrees to trade one of her magic broomsticks to Gaspar the ghost in exchange for one of his gold chains. Gaspar is somewhat skeptical that the broomstick is in working order and insists on a guarantee equal in days to the number of links in his gold chain. To facilitate enforcement of the guarantee, he insists on paying by the installment plan, one gold link per day until the end of the 63-day period, with the balance to be forfeit if the broomstick malfunctions during the guarantee period. Wanda agrees to this request, but insists that the installment payment be

effected by cutting no more than three links in the gold chain. Can this be done, and, if so, what links in the chain should be cut? The chain initially consists of 63 gold links arranged in a simple linear order (not closed into a circle).

Stop reading and try to determine an additional property possessed by the goal that would be helpful to derive for use in solving the problem.

The primary property of the goal that is useful in solving the problem is the information that only three links need be cut to achieve the goal. This property means that there will be at least three single links in the goal state, namely, the three links that have been cut. We still do not know how long the other lengths of chain will be (which single links in the 63-link chain will be cut), but we can then begin to work on the problem, knowing the lengths of three of the segments of chain in the solution of the problem (the three single-link chains). This problem also illustrates the subgoal method, and is discussed in this context in more detail in Chapter 6. If you cannot wait until Chapter 6 to check your solution to this problem, please turn ahead to pages 100–101.

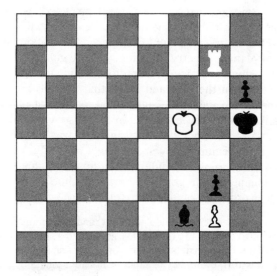

FIGURE 3-10
Part of a famous chess problem. White to move and to achieve mate in five moves.

In some problems, one cannot rigorously infer additional properties possessed by the goal, but one can make reasonable conjectures based on heuristic principles. For example, consider the following chess problem, which constitutes one portion of a famous problem originated by Sam Lloyd. The problem is for white to achieve checkmate in five moves from the starting position shown in Fig. 3-10. Stop reading and try to solve this problem by guessing one or more plausible goal positions in which black is checkmated and then try to determine how you could achieve such a checkmate position.

It seems reasonable to conjecture that the checkmate position will have white's rook at his own king's rook one. That is, white's rook will be at the end of the open file where he has the black king trapped. Black has the potential opportunity to interpose his bishop at two places in that file between the conjectured position of white's rook and black's king. However, the move sequences by which black can interpose his bishop between the white rook and the black king can all be frustrated by white in one way or another. The essential strategy for solving the chess problem comes by conjecturing that you wish to have the white rook at white's king's rook one, without the possibility of black blocking the attack by his bishop, and then working forward to determine what white must do at each move in order to achieve that terminal checkmate position.

4

Classification of Action Sequences

RANDOM TRIAL AND ERROR

The first thing that most people do when confronted with a problem is to start applying the allowable operations to the givens in the problem. Call this *random trial and error*. (Readers with a course in probability should understand that what I am calling random trial and error is equivalent to random sampling with replacement from the population of action sequences less than or equal to some maximum length.) If a very short sequence of such actions is sufficient to get from the givens to the goal, even randomly generated sequences of actions may yield the solution fairly quickly.

SYSTEMATIC TRIAL AND ERROR

To avoid going around in circles, it is obviously desirable to remember what sequences of actions have been tried already with no success. In addition, it is desirable to have a scheme for systematically generating different sequences of actions, which guarantees that all sequences (to some maximum length) will be generated. Most desirable of all random trial-and-error schemes would be a generation method that automatically produced a mutually exclusive and exhaustive listing of all sequences of actions up to some maximum length. Call this *systematic trial and error* (equivalent to random sampling without replacement).

From the above discussion, it should be relatively obvious that there can be different degrees of systematicness between random and completely systematic trial and error. A problem solver could have some memory for past attempts, but it could be limited or subject to error. A problem solver could also have different degrees of effectiveness in systematically generating all of the different sequences of actions. The degree of systematicness in the use of trial and error is one useful indicator of the intelligence of different species of animals. I think it is not known whether any species of animal below human beings can be trained to be more systematic in their trial and error, but humans certainly can be. People can overcome their memory limitations by writing things down, and they can often invent mutually exclusive and exhaustive generation schemes, though the difficulty of accomplishing the latter varies from problem to problem.

CLASSIFICATORY TRIAL AND ERROR

The most powerful kind of trial and error is what might be called *classificatory trial and error*, which requires that sequences of actions be organized into classes that are equivalent (or probably equivalent) with respect to the solution of the problem. That is, if one sequence of actions within a class will solve the problem, then all the other sequences of actions within the same class will probably also solve the problem. Conversely, if one sequence of actions within the class can be shown not to solve the problem, then probably every other sequence of actions in the same class will also fail.

To appreciate the power of classificatory trial and error, consider a state-action tree with n possible actions at each node of the tree. With this representation, there are n^m possible sequences of actions that are m actions in length. For even rather small values of n and m, n^m can be so large as to prohibit the use of systematic trial and error.

Obviously, if n^m sequences of actions could be reduced to a small number of *equivalence classes* — that is, classes that are equivalent with respect to the solution of the problem — it would make the problem much simpler to solve. In this case, you could systematically try one sequence from each class until you found a class that solved the problem. Such classificatory trial and error only works for problems where sequences of actions fall into classes that are equivalent with respect to solution of the problem, but most of the problems people solve probably exhibit some such equivalences.

There are four different categories of problems in which classificatory trial and error is helpful, and within each four are two subtypes. To discuss these types of problems, let us imagine that the states reached by various action sequences applied to the given information can be represented by a sequence of letters, where the first letter stands for the action taken at the first node, the second letter for action at the second node, and so on. Thus, abc represents the state reached by taking action a at the first node, followed by action b at the second node, followed by action c at the third node in the state-action tree. The basic principle is that two or more action sequences are equivalent if and only if they result in the same state or states thought to be equivalent with respect to solving the problem.

The first major type of equivalence class of *action sequences* is the obvious one that results from having equivalence classes of *actions*. In this case, for example, let us imagine that a set of actions $\{b_1, b_2, b_3, \ldots\} = \{b_i\}$ are all identical or thought to be equivalent. In this case, the action sequence ab_ic is equivalent to the action sequence ab_jc, for all i and j. People usually have no trouble in identifying equivalence classes of action sequences based on such elementary equivalences of component actions. Such equivalent actions often arise in problems where there are a large number of equivalent givens, such as a large number of entities of the same type—for example, six sticks identical in length and every other important property. When the givens are not identical in every respect but are equivalent with respect only to the properties that are thought to be important to the problem, then recognition of such equivalences may be more difficult and subject to error. In any event, identical or equivalent actions (often resulting from identical or equivalent givens) produce the first type of equivalence classes of action sequences.

A second and relatively familiar type of equivalence class of action sequences arises in problems having *commutative* actions—that is, where the result of taking action a followed by action b yields the same result as taking action b followed by action a. If three actions (abc) are all commutative with respect to one another, then action sequences abc, acb, bac, bca, cab, and cba are all equivalent, since they result in the same state when applied to the same given information or other starting point in a problem. For example, in solving for x, given the equation $5x + 17 = 3x + 21$, we could subtract $3x$ from both sides of the equation as the first step, then subtract 17 from both sides as the second step; but the same result is achieved by performing the actions in the reverse order. Even the final action of dividing both sides of the equation by 2 could be commuted with respect to the other two actions and equivalent results obtained.

A third major way in which action sequences may be equivalent occurs in problems where one or more actions have *inverse* actions. If action a has an inverse action a^{-1}, then the result of applying action a followed by action a^{-1} is to leave the state of the problem identical to what it was before the sequence aa^{-1}. Said another way, the sequence of actions aa^{-1} equals the identity action, which leaves the state of the problem unchanged. If all the actions in some sequence are commutative, then any co-occurrence of action a and its inverse a^{-1} permits you to cancel both a and a^{-1} from the sequence. For example, if the actions a, a^{-1}, b, c, d are all commutative with respect to each other, then the action sequence $ca^{-1}bda$ is equivalent to the action sequence cbd, since the a and a^{-1} cancel each other.

A good example of the power produced by the combined recognition of commutativity and inverse actions in reducing the number of different action sequences to a small set of equivalence classes is provided by the *six-arrow problem*:

> You are given six arrows in a row, the left three pointing up, and the right three pointing down. The goal is to transform these arrows into an alternating sequence such that the left-most arrow points up, the next arrow to it points down, the next up, then down, then up, and then down. The actions allowed are to simultaneously invert (turn upside down) any two adjacent arrows. Note that you cannot invert one arrow at a time but must invert two arrows at a time, and the two arrows must be adjacent. The given and goal states are illustrated in Fig. 4-1. Achieve the solution using the minimum number of actions (inversions of adjacent pairs).

Before reading further, try to solve this problem by determining the very small number of different equivalence classes of action sequences.

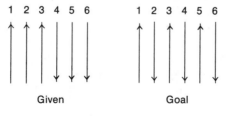

Given   Goal

FIGURE 4-1
The six-arrow problem.

If you have not solved the problem, consider the following. In representing the information given in the problem, you should note that there are only five different possible actions that you can take at any given stage of the problem — namely, to invert arrows 1 and 2, 2 and 3, 3 and 4, 4 and 5, or 5 and 6, as shown in Fig. 4-1. Of course, if every different sequence of actions had to be considered, and you had no way of knowing how long a possible sequence might be necessary to solve the problem, then the problem could be extraordinarily difficult, despite the limited number of actions available at each node. In fact, a little reasoning concerning the equivalence of different action sequences reduces the number of nonequivalent action sequences to an extremely small number.

In the first place, you should note that the order in which you perform the actions makes no difference. That is, the actions commute one with another, so that inverting arrows 3 and 4 and then inverting 4 and 5 is completely equivalent to first inverting 4 and 5 and then inverting 3 and 4. The same is true for any set of three or more actions in a sequence. Thus, you do not have to deal with ordered sets of actions but only with unordered sets. This statement simply means that all the different orderings (permutations) of a given unordered set of action sequences are equivalent, greatly reducing the number of possible solutions to be considered. Now stop reading and try to solve the problem, if you could not before.

If you still cannot solve the problem, consider the following hint. An optimal solution will contain no more than one occurrence of any given type of action. An action is its own inverse. Inverting arrows 2 and 3 twice leaves the arrows exactly the same as they were. Thus, any pair of two occurrences of a given action can be canceled (even if they are not adjacent, since the actions are completely commutative). Any even number of occurrences of an action is equivalent to zero occurrences of that action, and, for the same reason, any odd number of occurrences is equivalent to a single occurrence. Thus, we need consider only combinations (unordered sets) of from one to five possible actions. At this point we have reduced a potentially infinite number of different action sequences to 31 possible classes of action sequences. Each of the 31 classes can be represented by its simplest member, as follows: 5 single-step actions, 10 two-step action sequences, 10 three-step action sequences, 5 four-step action sequences, and 1 five-step action sequence. Stop reading and try to solve the problem, if you did not before.

We can now observe that two of the five actions — namely, inverting arrows 1 and 2 and inverting arrows 5 and 6 — cannot possibly be included in the optimal solution, since these actions change arrows 1 and 6, which are in the right position in the beginning state. To change these end arrows back to the correct position would require another use of exactly the same action, since no other actions change arrows 1 and 6. This solution cannot possibly be optimal, since it is equivalent to not performing the action at all. Thus, we have reduced the number of possible actions to consider at any node to three. The maximum number of actions in a solution sequence is now reduced to three.

It is then a simple matter to rule out all of the one-step and two-step action sequences, leaving only the single three-step action sequence as a solution to the problem. Of course, as illustrated in Fig. 4-2, there are actually six different action sequences that all achieve the goal in the smallest number of steps (three). These six solutions differ only in the order with which the three actions are applied. This solution points out once again the existence of a large variety of action sequences that are completely equivalent with respect to the solution of the problem.

The fourth major type of equivalence class of action sequences arises in problems where some arbitrary sequence of actions *abc* results in

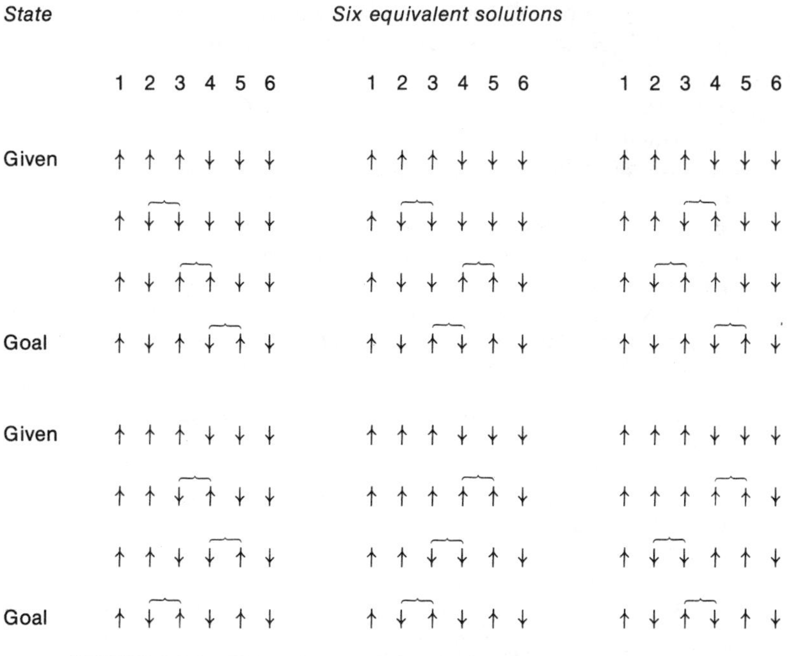

FIGURE 4-2
Six equivalent three-action sequences that solve the six-arrow problem.

an identical or equivalent state as some other sequence of actions *dafg*, for example. You may not have any elegant theoretical definition for all the equivalences of this fourth major type in a problem, but nevertheless you should recognize such equivalences and take advantage of them in speeding up the solution to the problem. In this fourth, most general case of classificatory trial and error, what we are doing is to define certain problem states and recognize that any action sequence that achieves a given problem state is a member of the same equivalence class. We may not know prior to executing an action sequence that it will result in the same state as some already executed action sequence; however, the advantage of recognizing the equivalent state reached by both action sequences is that we need not continue pursuing an action sequence *dafg* that arrives at the same state as a previously executed action sequence *abc*, if we have already determined that no path from that state is likely to reach the goal. Thus, we can truncate in this way, recognizing that a different sequence of actions has resulted in the same state, and, therefore, we need not continue the action sequence from that point on, since it would be, in essence, repeating sequences previously shown to be fruitless. One practical way to implement this use of classificatory trial and error is to give names or otherwise store in your memory a representation of certain distinctive states reached in various attempts to solve the problem.

A good example of the usefulness of explicitly identifying distinctive states achieved in various attempts at solution of a problem is provided by the *railroad-siding problem*:

> You are given a circular railroad track that passes through a tunnel and has one siding, as illustrated in Fig.4-3. In the given state of the problem, an engine, *E*, rests on the siding and two cars, *A* and *B*, rest on the circular track on opposite sides of the tunnel, as illustrated in Fig. 4-3. The goal is to interchange the positions of cars *A* and *B* and have the engine

(a) Given   (b) Goal

FIGURE 4-3
The railroad-siding problem.

back on the siding. An important restriction is that only the engine can pass through the tunnel; the cars cannot. Both cars and engine may rest on a siding in any order and in any numbers. However, as with real-world railroad sidings, a series of cars coming off the siding must go on to the circular track in the direction of the tunnel; they cannot make the sharp angle turn from the siding onto the circular track in the direction away from the tunnel. Cars can be coupled and uncoupled from one another or coupled or uncoupled from the engine at any point.

Stop reading and attempt to solve the problem by drawing a diagram on a piece of paper and getting three distinguishable objects to act as the engine and cars A and B.

You might consider there to be many distinguishable states in the problem, where the states are defined to be the different arrangements of the engine and cars A and B on the circular track and siding; however, it might be extremely useful to consider a simple distinguishable state in which only one car or engine rests in each of the three major portions of the track—namely, the siding, the portion of the track in a clockwise direction from the siding to the tunnel, and the portion of the track in a counterclockwise direction from the siding to the tunnel. Considering only these three different positions and limiting consideration to those cases where only a single car or engine rests in each of the three positions, there are only six possible configurations of the three entities in the three positions. One is the given state, another is the goal state, and the remaining four can easily be represented with pencil and paper. The six possible configurations result from the possibility of picking any of the three entities to fill a position on the siding, then picking any two of the remaining entities to fill the upper position on the circular track, which leaves only one remaining entity to fill the lower position on the circular track. This yields 3×2 or six possible configurations. Having listed all six configurations, you might find it useful to quickly classify your action sequences with respect to whether they achieve as a subgoal any one of the four configurations other than the given or goal configuration. All action sequences that achieve any particular one of the four nonterminal configurations are equivalent. Thus, you might set each of the four nonterminal configurations as a subgoal, try to achieve it, and then see whether you could get to the goal position from that particular nonterminal position. In this way, you ensure a certain degree of variety in the action sequences you take, so that you are not going around in circles. Now stop reading and try to solve the problem again, if you did not before.

In my opinion, the ideal subgoal position to work for is to interchange the engine and car A, placing the engine on the upper portion of the circular track and car A on the siding, with car B remaining in the lower position of the circular track. This subgoal configuration is probably optimal because it is just slightly more than halfway between the given state and the goal state in terms of the sequence of steps needed to solve the problem. Note that, although this immediate state is being called a subgoal state, there is no sense in which we have any reason to think that this state is closer to the goal than the given state. Thus, we are not really using the subgoal method, as this will be defined in Chapter 6. Rather, the basis for selecting this configuration as a state to work toward is simply that getting *to* it from the given state and getting *from* it to the goal state each represent equivalence classes of action sequences. Stop reading and try again to solve the problem, if you did not before.

Consider an extension of the method of classifying action sequences with respect to states achieved — namely, identifying action sequences in terms of the sequence of states achieved. In the present problem, it turns out that the solution sequence passes through five of the six simple configurations that we distinguished. The sequence of five simple configurations in the solution of the problem is shown in Fig. 4-4. By following this series of subgoals, you should be able to solve the problem. This problem provides an excellent illustration of the utility of identifying equivalent action sequences by states achieved or sequences achieved, since it is rather easy to get mixed up concerning where one is and where one is going with all the involved move-

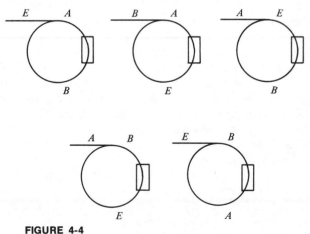

FIGURE 4-4
A sequence of simple configurations in the solution to the
railroad-siding problem.

ments of cars and engines required in order to solve the problem. If one does not identify landmarks along the way, it is easy to go around in circles.

So far, the examples that have been presented have illustrated only cases where several different action sequences were considered equivalent because they achieved precisely identical states. In some problems, it is useful to consider action sequences to be equivalent when they achieve states that are considered equivalent with respect to solution of the problem, despite the fact these states may not be identical in every respect. The states (and therefore the action sequences that lead to them) are considered equivalent because they all have certain properties in common that we judge to make the states equivalent insofar as solving this particular problem is concerned (though the states might well not be judged equivalent with respect to solving some other problem). Such classification of states (and action sequences) as equivalent is, of course, more dangerous than the completely safe classification of states as equivalent when the states are identical. If our judgment is faulty concerning which properties are relevant and irrelevant to the solution of the problem, then our judgment that all members of some equivalence class will fail to solve the problem may be faulty. Nevertheless, our judgment concerning relevant and irrelevant properties is generally sufficiently good that such equivalence classes are generally quite useful. An example of this type of equivalence classification of action sequences along with some of the identity-based equivalence classification of action sequences is provided by the *cheap-necklace problem*:

> You are given four separate pieces of chain that are each three links in length (see left side of Fig. 4-5). It costs 2¢ to open a link and 3¢ to close a link. All links are closed at the beginning of the problem. Your goal is to join all 12 links of chain into a single circle (see right side of figure) at a cost of no more than 15¢.

Stop reading and try to solve the problem by defining equivalence classes of action sequences based on the achievement of equivalent states.

If you did not solve the problem, consider these hints. There is an implicit operation of *inserting* one link into an open link that has a cost of 0¢ attached to it. In addition, there is another implicit operation of *detaching* an open link from a closed link that also has a cost of 0¢. These operations are only implicitly specified. Now try to solve the

Given state   Goal state

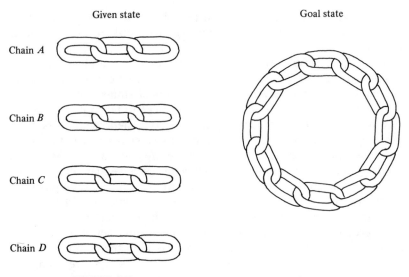

Chain A

Chain B

Chain C

Chain D

FIGURE 4-5
The given and goal states for the cheap-necklace problem.

problem again, if you did not before because you did not explicitly represent these operations. If that was not your difficulty, read on.

Having represented all the important givens and operations in the problem, let us examine how many different types of action sequences there might be that achieve the goal of getting all the links into a closed chain (circle). The one type of action sequence that virtually everyone considers first is to open an end link of one chain (for example, chain A), insert an end link of another chain into it (for example, chain B), close the joining link, open an end link of the combined (6-link) chain, insert another 3-link chain (for example, chain C) into it, close the joining link, open an end link of the combined (9-link) chain, insert the last 3-link chain (chain D) into it, close the joining link, open an end link of the combined (12-link) chain, insert the other end link, and close the joining link to form a closed chain. However, this action sequence costs 18¢, which exceeds the limit of 15¢.

There are a large number of action sequences that are essentially equivalent to the one just mentioned, which we might refer to as the *end-to-end* action sequence. Obviously, it makes no difference which 3-link chain we start with, add on second, or add on third. In addition, it makes no difference which end links we open at various stages of the problem. If we have explicitly noticed the equivalence of all these different action sequences, we are in a favorable position to discover whether there are any action sequences that are not equivalent to end-to-end that might result in the solution. If you have so far not solved the problem, stop reading and try again.

It might occur to you that, after opening a link, you could insert two end links into it or insert a middle link. These actions as a part of any action sequence (which did not later essentially reverse the effect of these actions) would undeniably result in an outcome that was not equivalent to end-to-end. However, a little inspection of the nature of the goal to be achieved reveals that, in the final closed chain, there are no links that have more than two links inserted in them. Thus, all methods that insert two end links (of chains with two or more links) or a middle link (of a chain with three or more links) into an end link of any chain with two or more links could not produce the goal state. (This statement is true unless the critical action were later reversed, and reversing actions seems very unlikely to be a part of the correct solution, in view of the cost limitation.)

Assuming that we have heuristically rejected all action sequences that involve actions that result in three or more links being inserted inside another link, are there any other types of action sequences that are not essentially equivalent to end-to-end? If you have not yet solved the problem, stop reading and try again, then read on.

Yes, there is exactly one other type, and it is the solution to the problem. For some reason, this type of action sequence does not occur to many people very quickly, but if you have ruled out all examples of the previously mentioned two equivalence classes of action sequences, then you are in a rather favorable position for discovering this remaining type of action sequence. At least, you are not wasting time trying out many specific examples of each of the two large classes of action sequences that do not work.

Perhaps you have already discovered this remaining type of action sequence and perhaps you have not. In any event, this type of action sequence could be called *destroying a chain*. The action sequence is as follows: Open a link of one (3-link) chain (for example, chain *A*), detach the link from that chain (chain *A*), insert the link into ends of *two* different 3-link chains (for example, *B* and *C*), close the joining link; open another link of the chain one is destroying (chain *A*), insert that link into an end of the combined (7-link) chain and an end of the remaining 3-link chain (*D*), close the joining link; open the last link of the first chain (*A*), insert the ends of the combined (11-link) chain into it, and close the joining link. The cost of this type of action sequence is exactly 15¢, solving the problem within the cost limitation.

MACROACTIONS

Now consider a sequence of actions that starts with some given state and achieves some other state. Call that a *microaction* sequence. Furthermore, consider any other sequence of microactions that starts from the same given state and achieves the same terminal state to be a member of an equivalence class of microaction sequences. Call such an equivalence class a *macroaction*. Thus, a macroaction is defined to be an equivalence class of sequences of microactions, though in some cases the equivalence class may consist of only one member.

Defining one or more macroactions based on the microactions specified in the problem is sometimes a significant aid in solution. If you like silly analogies, it is like wearing seven-league boots and taking giant steps instead of baby steps. Often, when one has defined macroactions, the number of such macroactions necessary to go from the givens to the goal is extremely small. This means a small state-action tree, with only a relatively small number of distinct possible macroaction sequences to test. Systematic trial and error will often be quite adequate to solve the problem from this point on.

Defining macroactions from sequences of microactions does have one possible difficulty, namely, that application of a macroaction might take one *past* the goal. In problems with destructive operations, one would have sped past the goal much as an express subway train speeds past a local subway stop. You would not even have the opportunity to see the goal for an instant as you passed through it, whereas at least on the express subway there is some chance that you could see your local stop as you sped through it.

Even with nondestructive operations, the effect of applying a macroaction that took you past the goal could be much the same as with destructive operations. To be sure, even when you have gone past the goal, you still have achieved it in some sense, when the problem involves nondestructive operations. However, if you do not know that the goal was achieved, because you never wrote down the goal expression, then in a practical sense, you have not achieved the goal.

As an example of the successful use of macroactions and classificatory trial and error, consider the set of possible action sequences involved in reducing an equation of the form $ax + b = cx + d$ to an equation of the form $x = \underline{\hspace{1cm}}$. There are a number of different microactions in different orders that will serve to reduce this equation. One could subtract b from both sides, then subtract cx from both sides, and then divide both sides by $(a - c)$, but one could also add $-cx$ to both sides, then multiply both sides by $1/(a - c)$, and then subtract $b/(a - c)$ from both sides, and so on. To someone experienced in solving such simple algebra problems, this problem may seem pretty trivial, but many students learning elementary algebra for the first time find these problems difficult. One reason for this difference is that the experienced linear-equation reducer sees all these different action sequences as equivalent, while many of the inexperienced linear-equation reducers have yet to learn this fact. Although probably few people actually think in terms of a single equation for linear-equation reduction, what an experienced linear-equation reducer has in his head is essentially equivalent to a single macroaction for problems of the above type, namely, $ax + b = cx + d \rightarrow x = (d - b)/(a - c)$. The experienced linear-equation reducer probably goes through a short sequence of microactions to solve such a problem, but such a person does this in one of a very small number of completely routinized ways, with, at most, a single choice of one of the equivalent microaction sequences. This statement is what is meant by thinking in terms of a single macroaction rather than a sequence of microactions.

Another example of the usefulness of thinking in terms of macroactions occurs in geometry construction problems. In such problems, you are allowed to use a compass, an unmarked straightedge (no gradations as on a ruler), and, of course, pencil and paper. The microactions that you have available are then to draw arcs of circles and straight-line segments. In learning how to solve geometry construction problems, you first learn what sequences of microactions allow you to achieve certain states, such as constructing a perpendicular to a line at a given point, constructing a perpendicular bisector of a line segment, constructing an angle bisector, or constructing a parallel to a given line through a given outside point. Thereafter, in more complex geometry construction problems, you think in terms of what sequence of these macroactions is necessary in order to solve these geometry construction problems rather than in terms of the original microactions of drawing arcs and circles and straight-line segments, though you must use a sequence of such microactions in achieving each macroaction. However, in constructing the basic plan for solving a more complex geometry construction problem, the use of a macroaction is extremely helpful. As an example of how thinking in terms of such macroactions simplifies geometry construction problems, consider the following problem, which was previously discussed briefly in Chapter 3.

Given an acute angle UVW and a point P within the angle, use a compass and straightedge to construct a segment QR passing through P, such that QP and PR stand in the ratio 2:1, Q and R lying on UV and VW, respectively.

Of course, the line QR shown in Fig. 4-6 is not part of the given state but is rather the goal to be achieved. The line QR has simply been drawn in the figure to facilitate thinking about the problem. Now stop reading and attempt to solve the problem, thinking in terms of geometry construction macroactions.

If you did not solve the problem, consider the following hint. The line segments QP and PR will be in the ratio 2:1, if and only if the ratio of the line segment QR to the line segment PR is in the ratio 3:1. This hint merely exemplifies a relatively trivial inference made from the given information, though in this case this trivial transformation of the given information can be of considerable help in solving the problem. Stop reading and try to solve the problem, if you did not before.

If you have not yet solved the problem, consider the following additional hint. One way to make the line segments QR and PR be in the ratio 3:1 is to make them corresponding parts of similar triangles, one of whose other sides is known to be in the ratio 3:1. Since PR is part of the line segment QR, the obvious choice for similar triangles would be to construct a parallel line to the line UV through the point P, producing a little triangle MPR (illustrated in Fig. 4-7), which would be similar to the big triangle VQR. Of course, we have not yet determined the line QR, so this operation is still to be done in order to solve the problem. However, any line QPR drawn through the point P will now result in triangle MPR being similar to triangle VQR. Thus, all that remains is to determine which line QPR will result in a similar triangle in which the ratios of the sides are 3:1.

Note that constructing a parallel to a given line through a given outside point is not an elementary microoperation, but rather a macro-

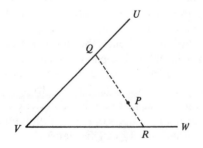

FIGURE 4-6
Construct QR such that $QP = 2 \times PR$.

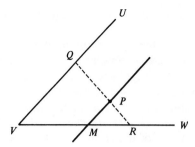

FIGURE 4-7
Constructing a parallel (PM) to UV
through the point P.

operation that requires a sequence of microoperations to be achieved. However, in planning the solution of the problem, we need not bother to explicitly carry out the sequence of microoperations necessary to achieve this macrooperation. Stop reading and try to solve the problem, if you have not done so.

Placing the line segments QR and PR in a 3:1 ratio is the same as placing the line segments VR and MR in the relation of 3:1. The latter is equivalent to placing the line segments VM and MR in the ratio 2:1. The length of VM is already determined. Therefore, all we need is to determine the length MR (which determines the point R). Since MR is half the length of VM, what we need is to determine what half the length of VM is. By this time it should be clear what macroaction will allow us to determine half of VM and mark it off from point M along the line MW to determine the point R. Stop reading and see if you can determine this macroaction and then solve the problem.

The required macroaction for determining half the line segment VM is to construct a perpendicular bisector to the line segment VM. This procedure determines the midpoint of VM, from which one can determine the length of half the segment VM by measuring from the midpoint to either point V or point M with the compass. Then we simply hold the compass at this position, place one end of the compass at point M, and mark the other point along the line MW to determine point R. Having determined point R such that the line VM is in the ratio 2:1 to the line segment MR, we have now uniquely determined the line QPR such that the ratio of the line segment QP to the line segment PR is 2:1, and the problem is solved.

For any readers without experience in plane geometry construction problems or who have forgotten what they learned, a bit of instruction concerning the achievement of the two principal macroactions used in this problem may be helpful in making the solution of the problem completely concrete.

To construct a parallel to the line UV through the point P, use the arbitrarily drawn line QPR in Fig. 4-6. Place one point of the compass at point Q and draw an arc through the lines QU and QP. The arc may be of any reasonable radius. Now draw an arc of the same radius around the point P. Now use the compass to measure the distance between intersections of the arc around Q that intersects QU and QP. To do this operation, place one point of the compass at the intersection of this arc with the line QU. Now, keeping the same radius with the compass, place one point of the compass at the intersection of the arc around the point P with the line PR and measure off the same distance along that arc. This point when connected to point P will produce a line parallel to the line VQU.

To achieve the macroaction of bisecting line segment VM, place the compass at point V and draw an arc around V intersecting the line VM at a point more than halfway between V and M. Now draw the same radius arc around the point M, intersecting the line VM and intersecting the arc around V once above the line and once below the line. Connecting the two intersecting points for the arc around V and the arc around M results in a perpendicular bisector to the line VM and therefore determines the midpoint of line VM.

Considering how complex especially the first of these two macroactions is, in terms of the sequence of required microactions, it is clear why it facilitates planning the solution of the problem to think in terms of macroactions rather than the sequences of microactions necessary to achieve them.

Knowing which sequences of microactions or equivalence classes of sequences of microactions to define as macroactions appears to depend very heavily (if not completely) on specific knowledge of the area from which the problem is taken. General problem-solving analysis makes it clear what the potential value is in defining macroactions, but it does not tell you which macroactions to define in any given problem area. This characteristic is very frequently the nature of the relationship between general problem-solving methods and specific knowledge — that is, general problem-solving methods direct you toward the type of specific knowledge that you should acquire and motivate you to acquire this knowledge by demonstrating their usefulness in the solution of problems.

Incidentally, there are other ways to solve this problem, using other geometric macroactions. So, if you thought of a different way to solve the problem, it may well be correct.

GETTING OUT OF LOOPS

Beginning problem solvers frequently run out of ideas to apply to a problem. Highly skilled problem solvers often experience the opposite difficulty; they have too many ideas and are forced to choose among a variety of possible approaches to the problem. I have devoted some space in this book to this question of deciding among many possible problem-solving methods, but I have been mainly concerned with the matter of providing the student with a rich variety of general methods for attacking problems to ensure that you do not spend a great deal of time staring at a problem without getting any ideas.

Sometimes you may have no ideas at all as to how to solve a problem, but more frequently you will run out of new ideas after having tried various methods, none of which worked. In such cases, you may repeatedly think of the inadequate methods for solving the problem and get the feeling that you are going around in circles. When you are caught in a loop like this, it is obviously time to do something different from what you have been doing. But how? In many cases that seems to be just the trouble: you are in a series of loops, thinking of the same inadequate ideas over and over again.

An excellent first step in getting out of a loop and doing something different is to analyze what you have been doing. You must determine the attributes (properties) of the approaches you have been taking. Usually when you make an effort to characterize what you have been doing in trying to solve a problem, you can immediately think of some ways to approach the problem differently. Often, what is critical is to step back and think about what you have been doing rather than think about the problem itself.

There are two basic levels at which this analysis of your problem-solving methods can take place: (a) the level of the specific action or action sequences specified in the problem (classifying action sequences) and (b) the level of general problem-solving methods (classifying problem-solving methods). In each case, after you have characterized what actions or methods you have used, you should ask what other classes of actions or methods seem remotely applicable to the problem.

At the level of general problem-solving methods, there are generally many specific ways to implement any given general method in any particular problem. What are the properties of the way you have chosen? Could you construct an alternative way that had different properties? Is there any information that is explicitly or implicitly a part of the problem that has not been explicitly represented? What kind of information has been used? Can you think of any alternative representation of this same information?

At the action-sequence level, a good example of the usefulness of classifying action sequences to get out of loops is provided by the *nine-dot four-line problem*:

> Without your pencil leaving the paper, draw four straight lines through the following three-by-three array of nine dots (see Fig. 4-8).

• • •

FIGURE 4-8
• • • The nine-dot four-line
problem.

• • •

Stop reading and try to solve this problem.

If you are like many amateur problem solvers, you may have produced a number of attempted solutions such as those shown in Fig. 4-9. Although there are many different ways you can produce incorrect solutions of this type, you can get the feeling rather quickly that they fall into a small number of classes, all of which are incorrect. You may feel you are going around in circles, producing attempted solutions that are of the same character as your previous tries and getting no closer to solution with each attempt. When you reach such a stage, it is well to try to determine the properties of your attempted methods of solution. If you ask what all these action sequences have in common, one answer is that they all keep the four lines within the perimeter of the three-by-three array of nine dots. If you examine the given information in the problem, it is clear that this restriction to the perimeter of the array is not a part of the problem. Thus, it is permissible to attempt solutions in which the lines extend beyond the perimeter of the array of dots, and, with this insight, the solution is readily achieved as illustrated in Fig. 4-10.

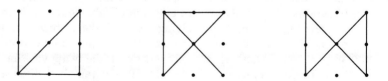

FIGURE 4-9
Incorrect solutions to the nine-dot four-line problem.

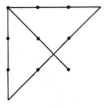

FIGURE 4-10
Correct solution to the
nine-dot four-line problem.

INCUBATION

When you have been going around in circles and wish to do something different to try to solve a problem, probably the most frequently given piece of advice is to put the problem aside for several minutes, hours, or days, and work on something else or get a good night's sleep before coming back to the problem. This is good advice, though in an examination situation the maximum period of time you can let any problem incubate is, of course, set by the time limitations of the exam. But even then it may be best to work on other problems and come back later to the more difficult ones, so that you will not spend too much time on difficult problems and fail to finish a number of easier questions. In addition, even a few minutes or tens of minutes spent solving other problems may give you a fresh perspective for solving problems you found difficult on the first attempt.

I must confess that incubation is not one of my favorite problem-solving methods, primarily, I suppose, because, when one is forced to use it, it indicates that all the other general problem-solving methods have failed. However, when you have tried a large number of approaches to a problem with no success, there comes a point at which even the most skilled problem solver should undoubtedly put the problem aside for a few hours or days and come back to it later. This is true even though a skilled problem solver may still be able to generate new ideas concerning how to solve the problem.

Psychologists do not understand why incubation is useful in solving problems. The difficulty in explaining the beneficial effects of incubation on problem solving is not that we lack any ideas concerning possible mechanisms for the effect. On the contrary, there are too many possible mechanisms for the beneficial effects of incubation on problem solving.

First, you may be quite generally fatigued after you have worked on a problem for a long time, and coming back to it in a fresher state of mind seems likely to be beneficial (though again we do not understand the mechanisms of general intellectual fatigue or the need for sleep and so on).

Second, there may be more specific intellectual fatigue or interference in the use of your memory because of the large number of incorrect actions you have taken in trying to solve the problem. The passage of time filled with intervening activities provides an opportunity for these interfering memories to fade away. Only the most valuable lessons you have learned remain in the foreground of your mind when you go back to the problem, with a host of lesser interfering associations having decayed to a low level. It is not clear that this sort of memory loss should necessarily be beneficial to problem solving, but it well might be.

Third, when you come back to the problem, you have an altered memory and new set of things on your mind as a result of the intervening activity. These new associations and new cues may well result in the retrieval of new ideas from memory concerning how to solve the given problem. This explanation is probably the single most plausible reason for the success of the method of incubation.

There is a fourth, somewhat more exotic possibility, namely, that a person's mind goes on unconsciously working the problem all during the long incubation period. Either because the unconscious mind has a long time to work on the problem or because something special is added by unconscious problem solving, the problem manages to get solved in this way, when conscious problem solving has failed. In any event, the unconscious problem solving may modify memory in a manner that facilitates conscious problem solving at a later time. There is not one shred of evidence for this explanation of incubation, whereas the first three possible mechanisms are all extensions of previously established psychological principles. Nevertheless, many psychologists believe in unconscious problem solving. I am very skeptical on the matter, but that is primarily a matter of philosophical preference.

In any event, incubation often works, whatever the mechanism.

5

State Evaluation
and Hill Climbing

THEORY

In the last chapter we reduced the amount of trial-and-error search in a problem by constructing equivalent state-action trees of reduced size. In this chapter, we discuss a very different way of reducing the number of state-action sequences that have to be searched before achieving the solution. The method has two parts: (a) defining an *evaluation function* over all states including the goal state and (b) choosing actions at any given state to achieve a next state with an evaluation closer to that of the goal. Picking an action on the basis of such a local evaluation of its consequences is known as *hill climbing*, since evaluation functions are frequently defined so that the goal state has the maximum value on some one-dimensional evaluation function.

Figure 5.1 illustrates the application of state evaluation and hill climbing to the state-action tree for some unknown problem with a hypothetical evaluation function defined over each state. The value of the function for each state is written inside the circle for each node (state). This example arbitrarily uses an integer-valued evaluation function, with the beginning state having value 0, the goal state having value 10, and nongoal states having values intermediate between 0

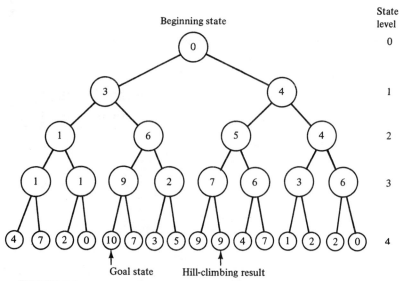

FIGURE 5-1
State-action tree with an integer-valued evaluation function defined over every state (node). One-step hill climbing results in the action sequence shown by the arrow. Note that, in this case, hill climbing does not achieve the goal state.

and 10. Application of a one-step hill-climbing method to this state-action tree with this evaluation function yields the sequence of action choices shown by arrows in Fig. 5-1. You will note that hill climbing need not succeed in achieving the goal the first time, and this time it did not.

Having failed to achieve the goal by hill climbing in the first attempt, there are many things you can do to achieve the goal, still using hill climbing. You could try choosing the action with the next to best value at one of the various nodes on the original hill-climbing path, use strict hill climbing at all other nodes, and see if you achieved the goal with any of these minimal violations of the general hill-climbing method. In the present instance, this minimal modification of hill climbing would succeed if you took the next to best action going from state level 0 to state level 1, because, from that point on, hill climbing results in an action sequence that achieves the goal.

Alternatively, you could try two-step hill climbing and choose the sequence of two actions at any given node that resulted in a node with the greatest value. This two-step hill climbing would produce the goal the first time in the problem shown in Fig. 5-1.

Finally, you could question the evaluation function you had defined over the states in the problem. There is usually no way to be certain that you have defined the evaluation function that is ideal for representing progress in achieving the goal in any given problem. Sometimes the failure of hill climbing suggests that a reexamination of the (explicit or implicit) evaluation function is in order. Evaluation functions are generally not *given* in the problem (except in *optimization* problems), and so any evaluation function can be chosen to see if it works in conjunction with hill climbing (or some other problem-solving method) to produce the solution to the problem.

Sometimes when hill climbing is used in conjunction with a state-evaluation function, a real-valued (numerical) evaluation is defined for each state. In other cases, you may have some ability to compare several states and judge which is closer to the goal, but no actual numbers are assigned to the states. Whether or not numbers are assigned to states, two states can have equivalent evaluation and so you could not choose between them.

So far we have discussed problems with only a single-valued (one-dimensional) state-evaluation function, but there are also problems where the goal differs from the beginning state on several dimensions. In these cases, it is usually possible to make judgments regarding closeness to the goal on each of the dimensions separately, but there may be no single, necessarily optimal way to combine the evaluations on each separate dimension into a single overall evaluation of each state. Thus, you could have a vector-valued evaluation function assigned to each state, as shown in Fig. 5-2.

There are a number of hill-climbing options in regard to vector-valued evaluation functions, such as that shown in Fig. 5-2. You could try various alternation schemes—that is, hill climbing on one dimension for a while and then hill climbing on another dimension for a while. Obviously, when no improvement is possible on a particular dimension by any action that you could take from the node where you are currently located, you should hill-climb on a different dimension for at least that node. If you have reached the goal with respect to one dimension, you should also hill climb on other dimensions. In using these alternation schemes, it helps to keep records of the nodes where you could have chosen to improve on a different dimension than the one you did choose. When the first hill-climbing path through the state-action tree fails to produce the solution, these nodes where you had good alternative choices are the obvious places to back up to and start new paths.

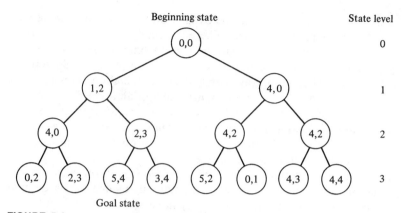

FIGURE 5-2

State-action tree with a two-dimensional vector-valued evaluation function defined over every state (node). In this case, the goal state has the evaluation vector (5, 4), and the beginning state has the evaluation vector (0, 0). The path taken by a hill climbing method depends on whether you hill climb on weighted summed components or try some alternation scheme. In the former case, the exact weighing of the two component values is also important in determining the path taken by hill climbing.

Another approach to multidimensional evaluation functions is to combine the values on the separate dimensions into a single overall value for each state. If there is some single most natural way to combine them, do it that way first; but remember that, no matter how natural the combination method might be, it could be the wrong way to combine the values on the different dimensions—that is, wrong for achieving the solution by one-dimensional hill climbing. If the originally chosen combination method fails to work, try some other method of combination, alternation schemes with the original multidimensional evaluation function, multistep hill climbing, defining a new evaluation function, or the like.

APPLICATIONS

Examples of the use of state-evaluation functions and hill climbing abound in problem solving. For instance, when you plan a trip across the country on a map, you initially examine roads that go in nearly the right direction. The right direction is the direction that reduces the distance between where you are and where you are going at the fastest rate. Of course, choosing the road at the beginning of a trip that goes closest to the right direction may prove to be a bad choice.

This road may eventually lead to a dead end or require you to go far out of the way to reach the goal. In addition, planning a trip on a map usually involves other considerations—speed, scenery, or other properties—besides finding the shortest road between the starting and ending points. These considerations place you in the position of doing hill climbing on a vector-valued evaluation function. Despite all these complications, experience suggests that hill climbing is a prominent method used in solving trip planning problems with a map.

Pencil-and-paper maze problems are rather similar to trip-planning problems on a map, and people frequently use hill climbing in an attempt to solve them. However, challenging maze problems are usually deliberately constructed to frustrate a hill-climbing approach. Maze problems frequently require nonoptimal choices at early and middle stages of the solution and may even require detours (increases in the distance from the goal, as measured by the most obvious evaluation function of physical distance). On the other hand, maze problems usually do not involve considerations of road speed or scenic beauty.

Defining an explicit evaluation function and employing hill climbing is also useful in solving the one-heavy-coin problem discussed in Chapter 3:

> You have a pile of 24 coins. Twenty-three of these coins have the same weight, and one is heavier. Your task is to determine which coin is heavier and to do so in the minimum number of weighings. You are given a beam balance (scale), which will compare the weight of any two sets of coins out of the total set of 24 coins.

A suitable evaluation function for solving this problem would be the number of coins whose classification as heavy or light is known. At the beginning of the problem, the value of the function is zero, since none of the 24 coins is known to be either heavy or light. In the goal state, the heavy-light classification of all 24 coins is known, so the value of the function is 24. Thus, a hill-climbing approach would choose an action at each node that maximized the number of coins whose heavy-light classification is known.

A very large number of alternative actions are present at each node. For example, at the first node, you might weigh any one of the coins against any two of the other coins. In general, you might weigh any set of m coins against any set of n coins, where $n + m \leq 24$. The number of different pairs of sets of m and n coins that satisfy the restriction that $m + n \leq 24$ is extremely large. However, the most elementary consideration of the previously mentioned evaluation function and the hill-climbing approach immediately rules out all actions that do not involve weighing two sets containing equal numbers of coins in the two pans of the beam balance. This exclusion reduces the number of alternative actions considerably.

Furthermore, using the method of defining equivalence classes of actions discussed in Chapter 4, note that, at the first node of the problem, you have no way to distinguish different subsets of i coins; thus, you must consider any two sets of i coins to be equivalent to each other (in their likelihood of containing the heavy coin). This consideration reduces the number of different alternative actions at the first node to 12 — that is, a set of 12 coins is weighed against a set of 12 coins, a set of 11 coins against another set of 11 coins, 10 against 10, and so on, or 1 against 1.

If you explicitly inquire which of these 12 alternative actions results in the greatest number of known coins following the first weighing, you should be led to select the optimal action at the first node — that is, to weigh a set of 8 coins against another set of 8 coins, since this maximally increases the value of the evaluation function from 0 known coins to 16 known coins following the first weighing, whatever the outcome of the first weighing.

The same sort of evaluation function and hill-climbing approach can be used to solve more complex coin-weighing problems, such as those involving two heavy coins or one coin that might be either heavier or lighter than the other coins. When the coins are classified into three or more categories (for example, heavy, medium, and light), then it may be useful to use as an evaluation function the number of coin-classification pairings (for example, coin 1 is heavy, coin 2 is medium, coin 3 is light) that have been ruled out.

In all of the coin-weighing problems, from the simplest to the most complex, keep in mind that, after a given weighing, the value of the evaluation function may be different for the different outcomes of the weighing. In such cases, the value of the evaluation function for a particular weighing is usually best considered to be the expected value of the evaluation function across all different outcomes, where the value of the evaluation function for each outcome is weighted by the probability of obtaining that outcome. Thus, after the first weighing of eight coins against eight coins in the previously mentioned one-heavy-coin problem, the optimal choice in the second weighing is either to weigh two coins against two coins or three coins against three coins. In either case, the three outcomes of the weighing (tilt left, balance, tilt right) are not equally likely, nor does each outcome result in an equivalent increase in the number of known coins. For example, with the three against three weighing (out of the eight remaining coins), the probability of their balancing evenly is $\frac{1}{4}$, while the probability of tilt left is $\frac{3}{8}$, and the probability of tilt right is $\frac{3}{8}$.

For simplicity, let us use as the evaluation function the number of *unknown* coins, where the goal state has a value of zero unknown coins. Thus, hill climbing, in this case, means attempting to minimize the value of the evaluation function. Using this evaluation function, the value of a balanced outcome in the three-against-three weighing is two remaining unknown coins, while the value of tilt left is 3 and the value of tilt right is also 3. The overall evaluation of the three-against-three weighing, then, is $(\frac{3}{8} \cdot 3) + (\frac{3}{8} \cdot 3) + (\frac{2}{8} \cdot 2) = \frac{22}{8} = 2\frac{3}{4}$.

The three-against-three weighing produces the minimum expected value on the evaluation function. This fact can be seen by computing the expected value for the other three plausible weighings — namely, one against one, two against two, and four against four. The two-against-two weighing is almost as good as the three-against-three weighing, by this evaluation function. The two-against-two weighing has an expected value of $(\frac{1}{4} \cdot 2) + (\frac{1}{4} \cdot 2) + (\frac{1}{2} \cdot 4) = 3$. The four-against-four weighing has an expected value of $(\frac{1}{2} \cdot 4) + (\frac{1}{2} \cdot 4) = 4$. The one-against-one weighing has the poorest expected value of all — namely, $(\frac{1}{8} \cdot 0) + (\frac{1}{8} \cdot 0) + (\frac{6}{8} \cdot 6) = 4\frac{1}{2}$.

In terms of achieving the goal of determining the one heavy coin out of 24 in the minimum number of weighings, either the three-against-three weighing or the two-against-two weighing is optimal on the second weighing. Thus, in this case, hill climbing is a successful problem-solving method, since it chooses one of the two actions that will lead to the goal with the minimum number of total actions (weighings).

Solving simple linear equations provides another example of the possibility of successful use of hill climbing in problem solving. Consider the linear equation $9x + 7 = 5x + 15$ as the given, with an expression of the form $x = $ _____ being the goal. The blank, _____, represents some currently unknown real number that constitutes the value of x in the solution to the equation.

Initially, we might define a four-valued vector evaluation function for this problem, consisting of the coefficients of the x and numerical terms on the left-hand side of the equation and the x and numerical terms on the right-hand side. For the linear equation above, then, the value of the evaluation function at the given state would be (9, 7, 5, 15). The value of the evaluation function for the goal state is (1, 0, 0, _____), where _____ again indicates that we do not currently know what real number is acceptable in this position. We might choose actions at each step designed to increase the number of terms of this

four-valued vector evaluation function that are in agreement with
the corresponding terms of the evaluation function for the goal. Thus,
if we subtract $5x$ from both sides of the equation, the evaluation
function is changed to (4, 7, 0, 15), which is known to disagree with
the evaluation function for the goal in only the first two positions (the
agreement of the value in the fourth position with the desired value in
the goal expression cannot be determined). Subsequently, subtracting
7 from both sides of the equation changes the evaluation function to
(4, 0, 0, 8), which disagrees with the goal expression in only one position (the first). Finally, dividing both sides of the equation by 4 has an
evaluation function (1, 0, 0, 2), which is known to disagree with the
evaluation function for the goal in zero positions. The state achieved
at this point that includes the expression $x = 2$ constitutes the solution
to the problem.

Rather than think of this at all in terms of a four-valued vector
evaluation function, we can simply think of the number of "bad"
terms in the expression. Initially there are three bad terms. After
subtracting $5x$ from both sides of the equation (obtaining $4x + 7 = 15$),
there are only two known bad terms. After subtracting 7 for both sides
(obtaining $4x = 8$), there is only one known bad term. Finally, after
dividing both sides of the equation by 4 (obtaining $x = 2$), there are
no bad terms, and the problem is solved.

It may be somewhat difficult for someone experienced in solving
such simple linear equations to imagine that anyone actually uses this
sort of evaluation function and hill climbing in order to solve so
simple a problem. However, this approach could be used, and, very
likely, many beginning algebra students unconsiously use just such
a method in solving their initial linear-equation problems.

The more experienced linear-equation solver very likely thinks of
the problem in terms of three subgoals, namely, getting all the x terms
on the left side of the equation, getting all the numerical terms on the
right side of the equation, and dividing through by the coefficient of
the x term. However, this subgoal method (to be described in detail
in the following chapter) uses the same sort of evaluation function as
used by the hill-climbing approach to linear-equation problems.

Once you are an experienced solver of linear equations you probably
never think of evaluation functions, subgoals, or hill climbing at all
but simply solve the problem using the same type of action sequence
you have used in solving other such problems—namely, subtract the
x term on the right-hand side of the equation from the x term on the
left-hand side of the equation, then subtract the numerical term on
the left-hand side of the equation from the numerical term on the

right-hand side of the equation, and finally divide through by the coefficient of the x term. (This problem-solving method, knowing how to solve a problem because you recognize its relationship to other problems you solved previously, will be discussed in Chapter 9.) Thus, there are many different problem-solving methods that can all lead to roughly the same sequence of actions in solving a simple linear-equation problem. This simple problem is discussed primarily to communicate what is meant by such concepts as evaluation functions, hill climbing, subgoals, relations between problems, and the like.

Furthermore, hill climbing is frequently used to solve more complex equations or sets of equations, using as an evaluation function some measure or measures of the discrepancy in form between some equation you have produced and the goal equation. Thus, in solving equations involving exponential terms with the unknown in the exponent, a solver often takes logs of both sides of the equation to increase the similarity of the resulting equation to the goal equation (since in the goal equation the unknown is not in the exponent). In solving differential equations, you can integrate to get rid of the differential terms, and in solving integral equations, you can solve for the integrals or else differentiate in order to get rid of integrals, and so on.

The six-arrow problem discussed in Chapter 4 to illustrate the power of noticing equivalence classes of action sequences provides a very good example of a problem in which you can define multiple—at least three—different evaluation functions. The three evaluation functions differ considerably in their effectiveness for a hill-climbing approach. Recall that the six-arrow problem is as follows:

> You are given six arrows in a row, the left three of which are pointing up and the right three of which are pointing down. The goal is to transform these arrows into an alternating sequence such that the left-most arrow points up, the next arrow to it points down, the next up, then down, then up, and then down. The actions allowed are to simultaneously invert (turn upside down) any two adjacent arrows. Note that you may not invert one arrow at a time but must invert two arrows at a time, and the two must be adjacent. The given and goal states are illustrated in Fig. 5-3. You are to achieve the solution using the minimum number of actions (inversions of adjacent pairs).

Stop reading and try to define three different evaluation functions that *might* be relevant to solving this problem by hill climbing, then read on.

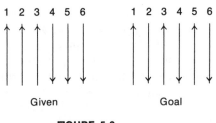

FIGURE 5-3
The six-arrow problem.

The most obvious evaluation function is probably the number of arrows that are in the same position as in the goal state. This evaluation function starts out at four in the given state and ends at six in the goal state. However, this most obvious evaluation function turns out to be of no help whatsoever in solving the problem at any of the early stages. For example, of the five alternative actions you might take at the beginning state, four leave the evaluation function unchanged at four and only one action—inverting arrows 3 and 4—decreases the evaluation (from four to two). Even this limited degree of discrimination among actions is of negative value in solving the problem, since inverting arrows 3 and 4 is an action that is, in fact, *desirable* to perform at some stage in solving the problem, whereas inverting arrows 1 and 2 and inverting arrows 5 and 6 are actions that should not be performed at any stage. Even if you choose to invert arrows 2 and 3 or invert arrows 4 and 5 at the first step, this evaluation function again provides no assistance in choosing the correct action at the second step. It is only when you have chosen the correct two beginning actions that the evaluation function could immediately tell you which action to choose at the third step, a fact that would be obvious in any event. Having read this discussion of one evaluation function, you might stop reading for a bit and try to generate some additional evaluation functions, if you are not satisfied with the ones you have thought of so far.

A somewhat different evaluation function that is considerably more useful in solving the problem is to count the number of *runs* of arrows (consecutive arrows with identical orientation). This evaluation function starts out at two runs for the beginning state and ends at six runs for the goal state. In the solution shown in Fig. 5-4, this evaluation function was not increased in going from the beginning state to the next state, but was increased at each of the two remaining states. In other solutions to the problem, the number of runs might be increased

State							Three evaluation functions		
							No. of arrows in goal position	No. of runs of arrows	Distance between two incorrect arrows
	1	2	3	4	5	6			
Beginning	↑	↑	↑	↓	↓	↓	4	2	3
	↑	↓	↓	↓	↓	↓	4	2	2
	↑	↓	↑	↑	↓	↓	4	4	1
Goal	↑	↓	↑	↓	↑	↓	6	6	

FIGURE 5-4
The values of three different evaluation functions for each successive state in a solution to the six-arrow problem.

at the first step, held constant at the second step, and finally increased again at the third step. Thus, the number of runs is a more useful evaluation function in conjunction with the hill-climbing approach than is the number of arrows in the goal position.

However, the evaluation function that is optimal in conjunction with the hill-climbing approach to this problem is to consider the distance between the two incorrectly placed arrows and attempt to reduce that distance. Probably you would arrive at such an evaluation function, in essence, by working backward (see Chapter 7) and noting that you could solve the problem if you had all the arrows correctly positioned, except two incorrectly positioned arrows that were adjacent to each other. In fact, you might note that, since an action always changes the position of two arrows, the final step must necessarily be to change two arrows (both of which are incorrectly positioned) to being correctly positioned. Thus, if more than one step is required, you know for certain that you could not get five arrows correctly positioned and be able to solve the problem. Hence, there is no point to the first evaluation function; you should focus instead on what you need to do in order to achieve the subgoal of putting the two incorrectly positioned arrows adjacent to each other. In essence, this procedure defines the third evaluation function, which is the distance between the two incorrectly positioned arrows. Note that, in the given state, the value of this evaluation function is 3, and the successive actions in a correct solution to the problem can reduce this to 2 and then to 1, from which the final action is obvious.

Another problem that illustrates the possibility of defining several plausible evaluation functions for the solution of a problem by hill climbing is the following *discrimination reversal problem*:

In the one-dimensional world of Lineland, there are two races of "people": whites and blacks. As in our three-dimensional world, the whites have for a very long time discriminated against the blacks. However, of late, the blacks have been making some gains in the area of social justice, in some cases obtaining judgments from courts and legislatures that a certain degree of reverse discrimination should obtain for a period of time, as symbolic retribution to blacks and as a lesson to whites concerning the evils of discrimination. One of the areas in which the blacks have just now achieved a court decision ordering discrimination reversal is in the matter of bus travel. In the past, whites have always ridden in the front of the bus and blacks in the back. Now the court has just ordered that for a time blacks will ride in the front and whites in the back.

When the order took effect, there was one seven-passenger bus that was already loaded with three blacks in the last three seats, three whites in the next three seats, and the front seat empty. The bus is automatic, requiring no driver (steering is not required in Lineland). All this is illustrated in Fig. 5-5. Since the order had already gone into effect, the police insisted that the blacks and the whites must reverse positions completely. Of course, people in Lineland are able to move to adjacent positions in their linear world, so a person could move to an empty adjacent seat in the bus. However, in addition, Linelanders have invented a special device that allows them to pass through two-dimensional space for a very limited distance, hopping over intervening persons and objects in either direction along their linear world. This hopping ability has a maximum limit equal to two seats in the bus. Thus, either a white or a black could jump over one or two adjacent seats in the bus, provided the target seat was empty. For example, in the given state of Fig. 5-5, the first white could move into the front seat, or the second white could hop over one white into the front seat, or the third white could hop over two whites into the front seat. But the first black could not hop over all three whites into the front seat. Using these movement properties of whites and blacks in

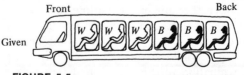

FIGURE 5-5
The discriminational reversal problem for a bus in Lineland.

Lineland, solve the problem of reversing the relative positions of the blacks and the whites so that all three blacks are in front of all three whites in the bus. Do this in the minimum number of moves. In solving the problem, note that it is irrelevant to the satisfaction of the court order where the empty seat occurs in the bus, so long as all three blacks are sitting in front of all three whites.

It is not possible in this problem to write down a single goal state, since a variety of possible goal states will satisfy the problem. All we can say for sure is that, in the goal state, the first four positions in the bus will contain all three blacks and the last four seats in the bus will contain all three whites, but one does not know where the empty seat will occur. This limited specification of the goal state, however, is quite adequate for defining a variety of evaluation functions that appear relevant to the solution of the problem. At this point, stop reading and, using a hill-climbing approach, try to define explicitly some evaluation functions that might prove useful for solving the problem.

There appear to me to be two obvious types of evaluation functions we can define for this problem, both of which are quite satisfactory for hill climbing. One evaluation function involves numbering the positions in the bus from 1 at the front to 7 at the back. The evaluation function would be something like the average position of the whites minus the average position of the blacks. We could then attempt to maximize this number. With this evaluation function, a detour is required on the first move, but thereafter all moves in the optimal sequence do increase this evaluation function by an amount that is either greater than or equal to every other alternative action (usually greater than every other alternative action). Explicit computation of the value of this function is somewhat more difficult than the second evaluation function to be discussed, but all that really counts is the relative difference between the current state and every alternative state that can be achieved by taking any admissible action. This difference is relatively easy to determine, and so this evaluation function proves quite helpful in conjunction with a hill-climbing approach to solving the problem. If you have not yet thought of a second evaluation function for the solution of this problem, stop reading and try to think of another one.

A second evaluation function, which is even easier to compute than the first, is the sum of the number of blacks in front of each white, summed across all whites. The value of this evaluation function for every state starting with the beginning state and ending with the goal state in the optimal solution of the problem is shown in Fig. 5-6. Clearly,

State	Configuration	Evaluation
Given (0)	☐ W W W B B B	0
1	W ☐ W W B B B	0
2	W B W W ☐ B B	2
3	W B W W B B ☐	2
4	W B W ☐ B B W	4
5	☐ B W W B B W	5
6	B ☐ W W B B W	5
7	B B W W ☐ B W	7
8	B B ☐ W W B W	7
Goal (9)	B B B W W ☐ W	9

FIGURE 5-6
The optimal solution to the discrimination reversal problem,
using hill climbing on the evaluation function of the number
of *B*s to the left of each *W*, summed over all three *W*s.

this evaluation function represents most closely what we are trying
to achieve in the problem. In the given state, there are zero blacks in
front of the whites, yielding an evaluation of zero. In the goal state,
there will be all three blacks in front of all three whites, yielding an
evaluation of 9. Thus, with this evaluation function, we know exactly
what value is possessed by the goal state.

With the first evaluation function, increases in the evaluation can be
made, after the goal is achieved, by moving whites further to the back
of the bus and positioning the empty seat in the middle. Since this
additional move is not required for the solution of the problem, it is
unnecessary and nonoptimal. However, this is a trivial matter, and, in
fact, both evaluation functions serve almost equally well in the solution
of the problem. Actually, the first evaluation function provides increases in the evaluation between states 2 and 3, states 5 and 6, and
states 7 and 8 (see Fig. 5-6), where the second evaluation function
cannot be changed by any action. Thus, in some ways, the first evaluation function is superior, though if you do some limited looking ahead
(two-step hill climbing) with the second evaluation function, you will
have the optimal choice at each of these three nodes.

The optimal solution to the discrimination reversal problem is
achieved by using (a) the second (relative position) evaluation function
with a limited degree of two-step hill climbing to decide among the
equivalently valued actions in states 0 and 1, 2 and 3, 5 and 6, and 7 and
8, or (b) the first (absolute position) evaluation function with the trivial
modification that you stop when the relative positions are correct,
even though the absolute-position evaluation function could still be

increased. It is a remarkable fact concerning the absolute-position evaluation function that it permits choice of the optimal action at each state, using a hill-climbing approach. That is, hill climbing using the absolute-position evaluation function will produce the solution the very first time by choosing the action that maximizes the evaluation on the next state (bearing in mind that, at the first move, you must choose the action that reduces the absolute position evaluation by the least amount).

The definition of evaluation functions and the use of hill climbing have a substantial role in playing chess games and in the solution of many chess problems. However, there are at least two major difficulties in the use of hill climbing in chess. First, even the immediate evaluation of any move you take must depend to some extent on your opponent's immediately following move. This fact leads to a certain degree of uncertainty, but it can be resolved in at least two ways: (a) by assigning subjective probabilities to your opponent's different moves and accordingly determining the expected values of your own possible moves (as was done in the "game against nature" illustrated by the coin-weighing problem) and (b) by assigning an evaluation to your move consistent with the best next move your opponent could produce (where this move is in any sense determinable). Since your subjective probabilities for your opponent's moves are likely to be only approximately accurate at best and since your ability to judge your opponent's best response is limited as well, there are substantial difficulties in applying hill climbing, no matter which approach is taken.

The second principal difficulty of using hill climbing in chess concerns the very large variety of different evaluation functions that are relevant to playing a good game of chess or solving many chess problems. For example, you are concerned with moving your own pieces to favorable positions on the board (where they control large numbers of squares), the subsidiary goal of control of the center, preventing your opponent's piece development, ensuring the safety of your king, jeopardizing the safety of your opponent's king, and many, many others. It is not at all obvious what all the relevant evaluation functions might be in chess nor how to weight them in different situations to come up with some overall evaluation of your next move.

Despite all these problems, the defining of evaluation functions and the use of hill climbing (looking ahead one or more steps) are important problem-solving methods in chess. To illustrate, consider the following very simple end-game problem involving black's rook and king against white's king, with the positions as illustrated in Fig. 5-7 and black to move. It is black's move and black's objective is to checkmate the

FIGURE 5-7
Black to move in a manner that maximally restricts
the squares of the board to which the white king
might eventually move.

white king in the minimum number of moves. Stop reading and try to
think of at least one evaluation function that is relevant to this objective
and that might dictate the choice for black's first move in the present
instance.

Although black must be continually concerned with avoiding stale-
mate (putting the white king in a position where he is not in check
but has no move except one that would put him in check), the most
obvious objective of black is to minimize the number of squares to
which the white king can move without being in check. Although minimizing the possible moves of the white king might be considered to
apply only to the next move, it is more useful to consider the evaluation
function to refer to minimizing the number of squares to which the
white king might *ever* be able to move (that is, minimizing the number
of squares reachable by a sequence of several moves). By the former
evaluation function, moving the black rook to either the fifth, sixth,
or seventh file (row) would be equally good. However, by the second,
more adequate evaluation function, only the move of black's rook to
his fifth file maximally restricts the number of squares on the board
to which the white king might ultimately move (if the rook merely
stayed on that file). Thus, this is the solution to the problem, and it is
relatively straightforward in terms of a hill-climbing approach using
the second evaluation function.

DIFFICULTIES WITH HILL CLIMBING

Local Maximum

One of the most common applications of hill climbing comes in optimization problems, where the evaluation function is already given in the first part of the problem. For example, suppose you are attempting to determine the maximum value of some function defined over a 1, 2, 3, . . . , n dimensional space. That is, you have some complex function of several variables for which it is possible to compute the value of the function, given any particular set of values for the individual variables, but for which it is not possible to determine by analytic means what the maximum of the function might be. To solve such problems, it is best to begin with a particular set of values for the independent variables, compute the value of the function (dependent variable) for that set of independent variables, then determine the value of the function for points that are nearby in the space of the independent variables. That is, you make small variations in the values of each of the independent variables in turn and determine the value of the dependent variable (evaluation function) for each new set of independent variables. Whatever direction of movement in the space of independent variables produces the greatest increase in the value of the dependent variable is chosen as the new focus for exploration. Proceed in this manner until you find a point for which no movement in any direction produces an increase in the evaluation function. At that point, you have climbed to the top of some local peak (local maximum) in the space, and hill climbing is no longer of any value in searching for the highest peak (absolute maximum) in the space of the independent variables.

The most frequently discussed difficulty with the hill-climbing approach in such optimization problems is that you can only reach a local maximum and have no guarantee that it is the absolute maximum (highest value of the dependent variable, defined over the space of the independent variables). The only real solution to the problem is to try a large number of widely dispersed starting points in the space of the independent variables and choose the maximum of the local maxima reached by hill climbing. Assuming that the givens of the problem do not include information concerning the optimum value of the evaluation function, you can never be absolutely sure that you have found the absolute maximum.

However, it should be pointed out that the application of hill climbing to most problems other than this class of optimization problems includes information concerning the value of the evalution function at the goal. In such cases, you can always know whether or not you have reached the goal by hill climbing. This greatly attenuates the seriousness of the "local maximum" difficulty.

End-bunching

Hill climbing is often used in construction problems, where you start putting together some of the materials to result in a state that is closer to the goal (more similar to the object being constructed) than was the original given state. In some construction problems, this method works well, but in others it is not useful. For example, consider the Instant Insanity problem described in Chapter 2. Stop reading and define some possible relevant evaluation functions for Instant Insanity.

One rather natural four-dimensional evaluation function might be the number of different colors you achieved on each side of the tower. The number of blocks already placed in the tower is included in this evaluation function in what appears to be a completely satisfactory way, since to achieve the goal of having four different colors on each side, you would have to have a tower of four blocks. If you were given more than four blocks to work with, this would not be a satisfactory evaluation function unless you considered the number of blocks, since you could achieve four different colors on each side by using more than four blocks. However, in the present problem, the goal state could be characterized exactly as having each of the four colors represented on each of the four vertical sides. Hence, we may consider the beginning state to have the evaluation vector $(0, 0, 0, 0)$ and the goal to have the evaluation vector $(4, 4, 4, 4)$. The four dimensions could rather naturally be combined into a one-dimensional evaluation function simply by summing the four components. In obtaining this sum, it seems natural to give equal weight to each component, since each dimension has the same range of values and an analogous meaning.

Although few people are explicitly aware of it, virtually everyone who works on Instant Insanity attempts to use some form of hill climbing, using something like the above evaluation function. Systematic use of this kind of hill climbing greatly reduces the search space (number of alternative towers to be investigated), but the method still leaves a very large number of alternatives to investigate. There are many equivalent options at each of the four nonterminal nodes of the state-action tree for Instant Insanity, so hill climbing with this evaluation function hardly yields the answer with a single series of four choices. The difficulty with this state evaluation function applied to this problem is that it is much harder to increase the evaluation function by the required amount at the last (fourth) choice node than at earlier nodes. At most of the last nodes, no action will achieve the goal, even though the solver is currently at a node that has the evaluation $(3, 3, 3, 3)$. Whether or not you can solve the problem is determined by the existence of such an action at the fourth node, but the evaluation function for the states that could be achieved at earlier nodes gives very inadequate information concerning the "correct" fourth node at which to be. That is, there are many fourth nodes with the evaluation $(3, 3, 3, 3)$, and very few of these have any action that leads to a terminal node with the evaluation $(4, 4, 4, 4)$.

There are many problems like this, where the restrictions bunch up at the end of the problem. It is as if you had many easy trails to climb most of the way up a mountain, but the summit was attainable from only a few of these trails, with the rest running into unscalable precipices. Hill climbing (in the problem-solving sense) is often not a very good method to use in such cases, though it may considerably reduce the amount of trial-and-error search.

The astute reader might note that the end-bunching of restrictions is a difficulty with hill climbing that is somewhat analogous to the local-maximum difficulty.

Detours and Circling

Problems with multiple equivalently valued paths at the early nodes can be difficult to solve with hill climbing, but perhaps the greatest frustration in using the method comes in *detour problems*, where at some node you must actually choose an action that decreases the evaluation. Somewhat less difficulty is encountered in what might be called *circling problems*, where at one or more nodes you must take actions that do not increase the evalutions. If the nodes where you must detour or circle have no better choices (that is, no choices that increase the evaluation), then you are more likely to try detouring or circling than if the critical nodes have better choices. When better choices are available, you tend to just choose them and go on without considering the possibility of detouring or circling. If the path you choose does not lead to the goal, you might go back and investigate alternative paths, but the first ones to be investigated will be those that were equivalent or almost equivalent at some previous node. Only after all of this fails should you try detouring—that is, choosing an action at some node that produces a state that has a lower evaluation than the previous state had.

The *missionaries-and-cannibals problem* is a famous example of the difficulties encountered by hill climbing in a detour problem. The problem is as follows:

> On one side of a river there are three missionaries and three cannibals. They have a boat on their side that is capable of carrying two people at a time across the river. The goal is to transport all six people across to the other side of the river. At no point can the cannibals on either side of the river outnumber the missionaries on that side of the river (or the cannibals would eat the outnumbered missionaries). This constraint only holds when there is at least one missionary on the side of the river where there are more cannibals. That is, it is all right to have one, two, or three cannibals on the same side of the river with zero missionaries, because then they would have no missionaries to eat.

Stop reading and try to solve the problem by explicitly defining some evaluation function and using a hill climbing approach, then see Fig. 5-8 for a sequence of states that solves the problem.

Offhand, you might think this was an absolutely trivial problem, since the state-action tree for the problem is rather small, and hill climbing on an evaluation function such as "the number of people on the other side of the river" reduces the number of paths to search to a very small number. But that is just the trouble! Hill climbing on this obvious evaluation function reduces the search space in such a way as to *eliminate* the path that leads to the goal. Given this evaluation function (the number of people on the other side of the river) for each state, there is a critical node at which you must detour (more than usual) to solve the problem by taking two people back across the river to the original side. Of course, at every other node in the problem, there is a necessary detour, when one person must row the boat back to the original side. But, as I mentioned before, necessary detours often cause little difficulty, especially in a problem like this one, where they are so obviously necessary on any path to the goal. But taking two people back to the original side is a detour that just does not occur to many people who work on this problem. If they were consciously aware that they had defined an evaluation function and were hill climbing using that evaluation function, then it would quickly occur to them that a detour might be necessary or that a new evaluation function was in order, and so on.

Incidentally, one reason why people have no difficulty with the necessary detour on every other node of the state-action tree for the missionaries-and-cannibals problem is that solvers usually automatically use two-step hill climbing; that is, they maximize the number

Node Level	State	Evaluation		Node Level	State	Evaluation
0	MMMCCC b	$\underline{0}$		6	MC ———— MMCC b	$\underline{2}$
1	MC ———— MMCC b	2		7	MMMC b ———— CC	4
2	C ———— MMMCC b	$\underline{1}$		8	MMM ———— CCC b	$\underline{3}$
3	CCC b ———— MMM	3		9	MMMCC b ———— C	5
4	CC ———— MMMC b	$\underline{2}$		10	MMMC ———— CC b	$\underline{4}$
5	MMCC b ———— MC	4		11	MMMCCC b	$\underline{6}$

(Critical detour step)

FIGURE 5-8
A diagram of the successive states in a solution to
the missionaries and cannibals problem, where M = missionary, C = cannibal,
b = boat, and the horizontal line is the river.
If two-step hill climbing is used until the last step
(to ignore the necessary detour on every alternate action), the evaluation
numbers considered are the underlined numbers.

of people on the other side of the river after a trip across the river
and the return trip as well. Using this two-step hill climbing, no detours
at all are necessary, and at the one crucial node, all that is necessary is
a circling action (not increasing the number of people on the other side
as a result of the round-trip voyage of the boat).

You could define an evaluation function different from "the number
of people on the goal side of the river." Obviously, you could use the
two-dimensional vector of the number of missionaries and the number
of cannibals on the goal side of the river, starting with (0,0) and the
goal being (3,3). This process does not avoid the necessity of detouring
(or circling in two-step hill climbing).

Somehow we would like to have the constraint regarding cannibals outnumbering missionaries to be reflected in the evaluation function. If that were done in the proper way, we would suspect that, by that evaluation function, it would not be necessary to make anything but the obviously necessary detour involved in getting the boat back to the original side of the river. If you were to consider the number of missionary-cannibal pairs on the goal side of the river to be your evaluation of the state, then the only detours that would have to be taken would be those necessary to bring the boat back across the river. This evaluation function does not distinguish between states as finely as does the previously mentioned evaluation functions. That is, at any given node, there are more actions with the same evaluation than is the case with the previously mentioned evaluation functions. However, this does not cause difficulties since most of the actions at any given node are eliminated from consideration by the constraint that the cannibals cannot outnumber the missionaries on either side of the river. I must say, though, that I am sure that someone who was thoroughly familiar with evaluation functions and hill climbing would look for a detour before defining some new evaluation function in this simple a problem.

Inference versus Action Problems

Most of the problems discussed in the present chapter as examples of the more or less successful use of hill climbing were *action* problems; only a few were *inference* problems. The best formal definition of the distinction between these two classes of problems is that action problems involve only destructive operations, whereas inference problems involve primarily or exclusively nondestructive operations. Action problems are concerned with achieving changes in some physical world via constructions, movements, or the like. By contrast, inference problems are concerned with our *knowledge* of something (whether or not one thinks of there being any physical referent). In inference problems, the objective is to expand the set of true statements to include the desired goal statement.

By this definition, trip-planning problems, maze problems, the six-arrow problem, the discrimination reversal problem, Instant Insanity, and the missionaries-and-cannibals problem are all action problems. The coin-weighing problem, the linear and other equation-solving problems, and the function-optimization problem are all inference problems (though other optimization problems might be action problems).

Obviously, the hill-climbing method is not restricted to action problems and excluded from inference problems. However, there is almost always a substantially greater economy (parsimony) when you describe your state at any point in the solution of an action problem. Since action problems involve only destructive operations, the state description can usually be achieved by a single simple expression. Furthermore, the complexity of the expression does not usually grow enormously with increases in the number of actions that have been taken in the attempt to solve the problem. By contrast, in inference problems the number of expressions generated increases with every action. In inference problems, you are continually increasing the number of statements known to be true. The description of the problem state must generally be considered to include the entire set of expressions given or derived up to that point. Since the goal is usually a single expression, it is generally much more difficult to define an evaluation function that is useful for hill climbing that compares the current state with the goal state.

Another reason for the greater difficulty in using hill climbing in inference problems is that the nondestructive operations frequently found in such problems are often not one-to-one operations—that is, operations that take one expression as input and produce one expression as output. There are such one-to-one operations, of course. However, in addition, inference problems usually contain a variety of two-to-one, and three-to-one, or even more complex operations—that is, operations that take two or three or more expressions as input and produce one expression as the output (the inferred expression). One-to-one operations are usually called unary operations; two-to-one and three-to-one operations are usually called binary and ternary operations. By and large, problems with only unary operations are more susceptible to a hill-climbing approach than are problems containing binary and ternary operations.

Of course, there is always the trivial evaluation function of how much knowledge you have obtained from the given information. However, sheer amount of knowledge (for example, the number of derived expressions), while positively correlated with achieving the goal expression, may not be very related to the achievement of the goal, if the inferences are proceeding in the wrong direction (a direction not related to inferring the goal expression).

We need to define an evaluation function that measures the relevant progress toward achieving the goal expression in inference problems, and such evaluation functions can frequently be found. However, when they are found, they are usually more useful in conjunction with the subgoal method (to be discussed in the following chapter) than they are with hill climbing. The reason is that usually a sequence of several actions is required to achieve an expression that moves the evaluation function in the direction of the evaluation characteristic of the goal. Most of the single actions taken in reaching each successive subgoal cannot themselves be identified as reducing the distance to the goal in terms of the goal evaluation function.

However, in these cases, if we define evaluation functions relevant for reaching each successive subgoal, then conceivably in a very large proportion of inference problems it is possible to use the hill-climbing method to achieve various subgoals. Some examples of this combined use of hill climbing and subgoal methods will be discussed in the mathematics, science, and engineering problems of Chapter 11.

6

Subgoals

THEORY

A problem-solving method that is important but difficult to master is that of defining subgoals in order to facilitate solving the original problem. This method is sometimes called "analyzing a problem into subproblems," or "breaking up a problem into parts." In essence, the purpose is to replace a single difficult problem with two or more simpler problems.

Of course, if you already know how to solve some of the subproblems, or if some of them are analogous to problems you already know how to solve, then obviously it might be easier to solve the set of simpler problems than the single original problem.

However, the fact that it is advantageous to break up a problem into subproblems does not mean you must be more familiar with the subproblems than with the original problem. One way to see the advantage of defining subgoals is to look at the following analysis of the state-action tree for a problem (1) with m alternative actions at each node and (2) a sequence of n actions being necessary for solution.

Let us assume that we know that the solution to some problem will require a sequence of n actions (or less). By systematic trial and error there are m^n alternative paths (action sequences) to be investigated in the original problem. Now assume that you can define a subgoal state that is known to be on the correct path to the goal and, let us say, halfway from the beginning to the goal. Defining one subgoal divides the problem into two subproblems — first, getting from the given state to the subgoal and, second, getting from the subgoal to the goal. In this case, there are $m^{n/2}$ paths to investigate in attempting to get from the givens to the subgoal, and there are the same number ($m^{n/2}$) of paths to investigate to get from the subgoal to the goal. Thus with the single subgoal, the number of action sequences to be investigated is $2m^{n/2}$ action sequences that are $n/2$ steps long, versus m^n action sequences that are n steps long in the original problem without a subgoal.

To get some concrete notion of the advantages of reducing the exponent of m in this manner, consider the case where $m = 10$ and $n = 10$. In this case, $m^n = 10^{10}$ and $2m^{n/2} = 2 \cdot (10^5)$. In this case, a single subgoal has reduced the search by a factor of 50,000, which is, of course, a staggering reduction. In addition, with the subgoal, the action sequences are only half as long. A state-action tree of a very simple problem, which vastly underestimates the power of the subgoal method, is shown in Fig. 6-1.

If we defined four subgoals (five subproblems) in the problem, with $m = 10$ and $n = 10$, then the number of two-step paths involved in achieving all the subgoals plus the final goal is $5(10^2) = 500$, which is a reduction of the search by a factor of $2 \cdot (10^7)$, or 20,000,000.

To be sure, a number of simplifying assumptions were made in computing the comparative advantages of defining a series of subgoals. However, the primary assumption, which overestimates the advantages of the subgoal method, is that you could be sure the subgoals you defined were states on a path that led to the goal. In some cases, you *can* be sure of this, but in many other cases you cannot. Nevertheless, if you could find a true subgoal by making 5, 10, or even 100 guesses of it, you would still be reducing the search space by extremely large factors in all of the many problems that require more than a few steps to solve.

The subgoal method is advantageous for attacking problems that require a sequence of more than two or three actions to solve — which is what most nontrivial problems require. Still, some problems are not simplified appreciably by this method; they are sometimes called insight problems because they require few steps to solve once the critical insight has been achieved. These include problems in which one must represent the components of the problem in some suitable way, guess the correct set of givens (where there are multiple given

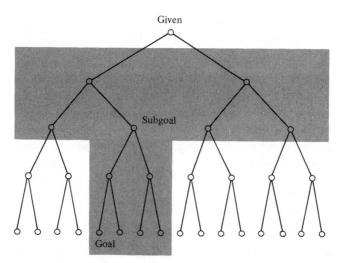

FIGURE 6-1
State-action tree for very simple problem showing how defining
a subgoal on the correct path (action sequence) to the goal can
reduce the search. In this case, the search is limited to the region
inside the two boxes, which is eight action sequences each two
steps long, instead of 16 action sequences each four steps long.
Some simplifying assumptions are made, such as that one knows
that the subgoal is two steps from the beginning and two steps
from the end. However, the average problem is much longer, and
the degree of reduction in search by defining subgoals is far
greater than in this simple example.

states), or choose a solution approach that violates hill climbing but
that requires choosing from among only a small number of action sequences, once the insight has been achieved. Examples of insight
problems, for which the subgoal method has little to offer, are many
implicit information problems, such as the notched-checkerboard and
block-cutting problems of Chapter 3, and the detour problems, such
as the missionaries-and-cannibals problem of Chapter 5.

The subgoal method also does not work, of course, if you cannot
think of any plausible subgoals. However, if the problem seems likely
to be a multistep rather than insight problem, it is usually advantageous
to spend some time trying to generate plausible subgoals, because of
the enormous power of the method.

How do you try to define a subgoal with reasonable hopes that you
are on a path to the goal? Although there is no method of defining
plausible subgoals that is mathematically precise and that applies to
every type of problem, you can take the first step by defining an

evaluation function over different problem states, as was done as a necessary precondition in applying the hill-climbing method. Having done this, you can recognize a plausible subgoal as a problem state with an evaluation part way between the given state and the goal state. If the evaluation function is multidimensional, then such a subgoal might have the goal values on some, but not all, of the dimensions. Or it might just have values *closer* to the goal values on some, or even all, of the dimensions.

Defining an evaluation function over problem states provides not only a way to *recognize* a plausible subgoal but also a way to *generate* or define plausible subgoals. Evaluation functions are single valued in the direction from a problem state to an evaluation vector for the problem state. (Multidimensional evaluation functions are still single valued so long as only one distinct evaluation vector is associated with each distinct problem state.) The inverse function may be multivalued, but that does not seriously reduce the value of this approach to generating plausible subgoals.

It may be that subgoals can always be determined to have intermediate values between the given and goal state according to some explicitly defined evaluation function. However, a person may frequently define subgoals without being able to state explicitly any relevant evaluation function. Thus, this book will generally discuss the application of the subgoal method to problems without attempting to describe any formal evaluation function.

In general, there are a multiplicity of plausible subgoals, some on a correct path to the goal and some not. As mentioned, even if the probability of a plausible subgoal being a true subgoal is only 0.1 or 0.01, the method is still reducing the search by an enormous factor in most problems. In any event, it is usually not very difficult to conjecture a variety of reasonably plausible subgoals, but the likelihood of defining a good subgoal will depend upon how good an evaluation function you have defined over problem states. In turn, how good your evaluation function is, how suitable it is to solving the problem at hand, often depends upon how adequately you represented the information in the problem (discussed in Chapters 3 and 10), the defining of macroactions, and the use of various other problem-solving methods. Problem-solving methods are generally used in combination, and the combined power of several methods in reducing the search space can result in very fast solution of many problems with little trial-and-error search being required.

When you have defined two or more subgoals to be achieved in getting from the given state to the goal, you can make a logical distinction as to whether the subgoals must necessarily be achieved in a certain order or whether they can be achieved in any order. This simple logical distinction between ordered and unordered subgoals is illustrated in Fig. 6-2.

In some problems it is obvious that one of the subgoals (SG_1) is closer to the given in terms of the evaluation function than is the other subgoal (SG_2), while the latter subgoal is closer to the goal than is the former subgoal. In such instances, the subgoals clearly should be achieved in a particular order.

In other cases, while the achievement of two or more subgoals may constitute two components that are necessary in order to get to the goal from the givens, it is not obvious which subgoal is easier to achieve from the givens. In the latter case, you have a choice of what order to arrange the subgoals on a path from the givens to the goal. In these cases, if your first choice is not working out well, then you should switch to some other choice in ordering the subgoals. In some cases, where the ordering of subgoals is not immediately apparent at the outset of the problem, some orders of achieving subgoals may be easier to accomplish than others. Being aware of the distinction between ordered and unordered subgoals permits greater flexibility in the solution of problems involving unordered subgoals.

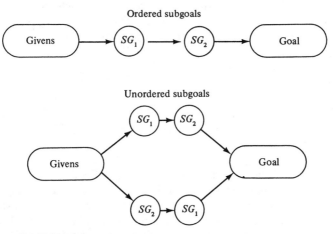

FIGURE 6-2
An example of ordered subgoals (SG_i) where there is a unique order in which the different subgoals must be achieved in getting from the givens to the goal versus an example of unordered subgoals where any order of achievement of the subgoals can lead to the goal.

If you have defined *n* subgoals (whether ordered or unordered), you have automatically defined *n* + 1 subproblems to be solved—namely, getting from the givens to one of the subgoals, getting from the first subgoal to the second subgoal, and so on, from the *n*th subgoal to the goal. Whether any particular ordering of the subgoals is required by the original evaluation function or is optional by that evaluation function, you may often be free to choose to work on any link in the chain first, second, and so on. By and large, it is advantageous to start with a subproblem of getting from the givens to one of the subgoals, or else to work on the problem of getting from one of the subgoals to the goal, because subgoals are frequently not fully defined problem states. It is usually preferable to work on a subproblem in which either the beginning or end state is completely specified. In most problems, this means starting from the givens is preferable, though in some cases working backward from the goal may be just as good or better. An additional advantage of starting with the givens is that you are simultaneously drawing inferences (see discussion of this method in Chapter 3) that you know is expanding the information you have available for the solution of the problem, no matter what the ultimate success of the particular subgoal approach that you have taken. If you begin working on other subproblems that do not start from the given state, then a failure using the subgoal approach may not have generated equivalently useful information as that which would have been generated by starting from the givens. All this is perhaps relatively obvious, but it is nevertheless important to bear in mind.

APPLICATIONS

Planning a trip across the country is a problem to which the subgoal method can be applied in a relatively trivial manner. If you wanted to travel from San Francisco to New York City, you might select Denver and Chicago as subgoals. Selection of such reasonable subgoals depends upon having an evaluation function defined over such cities (such as their two-dimensional coordinates on a map), which indicates that Denver and Chicago have intermediate values on the east-west coordinate compared to San Francisco and New York. In this case, it is primarily one of the two dimensions of the evaluation function that needs to be altered to get from the given state to the goal. But in going from Spokane, Washington, to Miami, Florida, by way of Denver and Memphis, Tennessee, you are substantially altering the values on both dimensions in going from the given state to the goal. Obviously, the subgoals are ordered in these trip-planning problems.

As an example of a problem that is extremely easy to solve using the subgoal method (with unordered subgoals), consider the following:

A light plane carrying three men crashes in the desert. The men decide that their best chance for survival consists of each of them setting across the desert in different directions in hopes that one of the directions will pass by a sufficient number of oases to permit that man to reach civilization and get help for the others. Before going their separate ways across the desert, they are faced with the problem of achieving an equal division of their stock of water and canteens. They have in their possession five canteens full of water, five canteens half-full of water, and five empty canteens. All canteens are the same size. Since water-carrying capacity is important should a man reach an oasis, they wish to divide both their supply of water and the number of canteens equally among themselves. How can they achieve this?

Stop reading and attempt to solve this problem.

If you were unable to solve it, consider the following hints. You might originally define three unordered subgoals consisting of attempting to divide the full canteens evenly among the three men, dividing the half-full canteens evenly among the three men, and dividing the empty canteens evenly among the three. It is immediately obvious that this subgoal approach will not work. An alternative definition of subgoals involves first making the inferences that the total quantity of water is $5 + \frac{5}{2}$ or $7\frac{1}{2}$ canteens full of water and that the total number of canteens equals 15. From this you can conclude that, in the goal state, each person will have $2\frac{1}{2}$ canteens full of water and 5 canteens. In essence, this defines a six-dimensional vector evaluation function such that in the given state each of the three persons has zero water and zero canteens and in the goal state each person has $2\frac{1}{2}$ canteens full of water and 5 canteens. If you have not solved the problem, stop reading and attempt to define relevant subgoals.

The relevant subgoals are to attempt to give first one of the men (it obviously does not matter which one) $2\frac{1}{2}$ canteens full of water distributed among five canteens. There are a number of ways of doing this, only one of which will make it impossible to achieve the other two subgoals of giving $2\frac{1}{2}$ canteens full of water in 5 canteens to each of the other two men. The only way that prevents achievement of the goal is to give the first man the entire set of half-full canteens. Any other method of giving the first man $2\frac{1}{2}$ canteens full of water and 5 canteens will permit achievement of the remaining two subgoals—namely, give the first man 1 full canteen of water, 3 half full canteens of water, and 1 empty canteen, or give the first man 2 full canteens of water, 1 half full canteen of water, and 2 empty canteens. Once the first subgoal has been achieved, it is trivially obvious whether or not it is possible to achieve the second and third subgoals.

Another problem that illustrates the power of the repeated use of the subgoal method is the following:

> Nine men and two boys want to cross a river, using an inflatable raft that will carry either one man or the two boys. How many times must the boat cross the river in order to accomplish this goal? (A round trip equals two crossings.)

Stop reading and try to solve the problem, using the subgoal method.

Define as a subgoal the problem of getting one man across the river and getting the boat back to the starting side. Stop reading and attempt to solve the problem, if you have not already.

It takes exactly four crossings to get one man across the river and return the boat to the original side. First, the two boys cross the river in the boat, then one boy takes the boat back to the original side of the river, then a man takes the boat across the river, and then the second boy takes the boat back to the original side of the river. These four crossings put both boys and the boat in the same position they were when they transported the first man across the river. Thus, to transport all nine men across the river will require 9×4 or 36 one-way crossings. At that point, both boys will be on the original side of the river with the boat, and one additional crossing will be required for them to get to the goal side of the river with the boat. Thus, a total of 37 one-way crossings are required in all.

An example of a somewhat similar subgoal problem in a probability context is provided by the following example:

> The ace, 2, 3, 4, 5, 6, 7, and 8 of hearts are placed face up in a row on the table. Then a pack of eight cards containing the ace, 2, 3, 4, 5, 6, 7, and 8 of spades are shuffled and placed in front of the player. As each successive spade is turned over, the corresponding heart is removed from the row. What is the probability that all the hearts can be removed without a break (hole) ever occurring in the row of hearts?

Stop reading and try to solve the problem.

Consider as a subgoal the probability that the first removed heart does not cause a break in the row. Stop reading and try to solve the problem, using this subgoal.

The probability of achieving the first subgoal (removing one heart without producing a break in the row) is exactly $\frac{2}{8}$. This probability results from the fact that there are two end positions to the row, and only the two cards in these end positions may be removed without causing a break. Since there are eight cards in the row, the probability is $\frac{2}{8}$. If you did not solve the problem before, stop reading and try to solve the original problem, having achieved the solution to the subgoal.

Once the first subgoal has been achieved of drawing one card from the end of each row (and its probability has been determined), the second subgoal should be to compute the probability that the second card will be removed from the end of the row. If the first subgoal has been successfully achieved, there are still two cards at the ends, but now only seven cards in toto. Thus, the probability of successfully removing a second card from the ends of the row is $\frac{2}{7}$. If you have not yet solved the entire problem, stop reading and try to complete the rest of the solution on your own.

Continue in this way, defining successive subgoals of removing cards from either end of the row until all cards have been removed. The probability that each subgoal will be successfully achieved with a random shuffling of the pack of spades is evidently $\frac{2}{8} \cdot \frac{2}{7} \cdot \frac{2}{6} \cdot \frac{2}{5} \cdot \frac{2}{4} \cdot \frac{2}{3} \cdot \frac{2}{2}$, or $\frac{1}{315}$. The probability of successful achievement of the entire set of necessary subgoals is simply the product of the probabilities of achieving each successive subgoal. Note that this probability problem represents a rather interesting variation in the use of the subgoal method, since there are, in essence, two parallel sets of subgoals involved in the problem. On the one hand, there is the series of subgoals of removing cards from an end of the row, progressively reducing the row in length without creating a hole. On the other hand, there is the series of subgoals of computing the probabilities of achieving each of these subgoals.

Now consider this rather different problem (previously discussed in Chapter 3), which also illustrates the use of the subgoal method:

Wanda the witch agrees to trade one of her magic broomsticks to Gaspar the ghost in exchange for one of his gold chains. Gaspar is somewhat skeptical that the broomstick is in working order and insists on a guarantee equal to the number of links in the gold chain. As a guarantee, he insists on paying by the installment plan, one gold link per day until the end of the 63-day period, with the balance to be forfeit if the broomstick malfunctions during the guarantee period. Wanda agrees to this arrangement, but insists that the installment payment be effected by cutting no more than three links in the gold chain. Can this cutting be done, and, if so, what links in the chain should be cut? The chain initially consists of 63 closed gold links arranged in a simple linear order (not closed into a circle).

Stop reading and try to solve the problem.

Assume it is possible to solve the problem by making only three cuts. Obviously, if it is possible, then Gaspar and Wanda will have to make change on various days during the 63-day period. That is, they must exchange various links of chain on the different days, so that Wanda acquires one extra link each day, since it surely is not possible to separate the chain into 63 individual links by making only three cuts. If you have not solved the problem, stop reading and try again.

Still assuming that it is possible to solve the problem, note that, if it is possible, the solution will result in creating at least three single links of chain, as well as various other longer links of chain. If you have not solved the problem so far, stop reading and try again.

Having three individual chain links will permit payment of one link per day from days 1 to 3. Now as the first subproblem, you should determine the longest link chain that can be used with Wanda making change, in order to permit payment of an additional link on the fourth day. Obviously, the solution to this problem is to cut a chain that is four links long, since Wanda can return the three individual links. Then the second problem is to cut the maximum link chain that will permit payment when the three individual chain links and the four-link chain have been used up. Obviously, this chain would consist of eight links. Continue in this manner, defining as subgoals the making of change, using lengths of chain known to be part of the solution, *until* these are all given over to Wanda and a longer chain is required. Then determine what longer chain is required on that day. Thus, the solution of the problem is to have 3 individual links of chain, then a chain each of 4 links, 8 links, 16 links, and 32 links. Since, by inspection, this will require only three cuts (separating the 4 from the 8, the 8 from the 16, and the 16 from the 32), the problem is solved.

Note that essential insight for solving the problem is to consider how to make change on each day of the 63-day period, starting from the first day and continuing through to the 63rd day, achieving these subgoals in order.

In addition to the subgoal method, it is also important to the solution of this problem for you to make the inference from the goal that, when you have determined where to place the cuts in creating the larger chain links, you will have also achieved three individual links.

Thus, you should start the process of change making on the initial days, using the individual links until they are inadequate, and thereafter continue to use all the known length chains until they are inadequate, at that point cutting off the largest length of chain that will solve the problem on that day. In a sense, the original problem is divided into 63 subgoals—that is, making correct change on each of the 63 days, though only a few of these days are special in that they require you to exchange one long piece of chain for all previously given shorter pieces of chain.

A very simple puzzle problem illustrates the use of the subgoal method in an entirely different context:

> Five squares are inserted into a three-by-two rectangle, as illustrated in the given of Fig. 6-3. Three of the squares have a label *A*, one square has label *B*, and one square has label *C*. Any square may be moved within the rectangle to an adjacent square, provided that the square moved into is empty. The problem is to make a sequence of moves so as to achieve the goal state, as illustrated in Fig. 6-3.

FIGURE 6-3
The given and goal states for the *ABC* puzzle.

Now make up five little squares of paper (or other tokens) that will fit in the rectangle in Fig. 6-3, and, by moving them around in the rectangle, attempt to solve the problem. In attempting to solve the problem, you will find it helpful to try to define a subgoal state that is on the path from the givens to the goal by some evaluation function.

Moving the *ABC* squares in the right four cells of the rectangle will not solve the problem, for the three squares can only be moved in a cyclic manner within the four squares, which will never change the relative ordering of the *B* and *C* squares—precisely what is required in the goal state. At some point, the *B* and *C* squares must be separated to the first and third columns of the rectangle in order to achieve a change in the cyclic order of the *B* and *C* squares. With this somewhat vague idea for a subgoal in mind, stop reading, attempt to define the subgoal more precisely, and then solve the problem, if you have not done so already.

A more specific definition of the subgoal of separating the *B* and *C* squares to opposite sides of the rectangle is illustrated in Fig. 6-4. Note that if the *B* and *C* squares are separated as in subgoal 1, it is relatively easy to move *B* next to *C* (subgoal 2) and then move the *A*'s around in such a way that *B* could be on top of *C* in the third column. Thus, in the case of this particular subgoal, you can quickly verify that you could get from the subgoals to the goal and then attempt to reach subgoal 1 from the given state. The latter problem is not too difficult and is left to you as an exercise.

Subgoal 1   Subgoal 2

FIGURE 6-4
A useful set of two subgoals for the solution of the *ABC* puzzle.

The relevant evaluation function for defining subgoal 2 is probably whether the cyclic order of *B* and *C* within the right four cells of the rectangle is *BC* or *CB*. Subgoal 2 shares the *BC* order with the goal state, whereas the cyclic order in the given state is *CB*. Since the *BC* ordering illustrated in subgoal 2 cannot be achieved from the prior state by moving *B* unless *B* is at some time moved out of the right four squares, you know that the preceding subgoal must have *B* in the extreme left-hand column of the rectangle. Subgoal 1 illustrates the simplest such possibility in relation to subgoal 2.

Surely one of the most remarkable simple examples of the use of the subgoal method comes in the *Tower of Hanoi (disk transfer) problem.* One version of the problem can be stated as follows:

> There are three identical spikes and six disks, each with a different diameter but each having a hole in the center large enough for a spike to go through. At the beginning of the problem, the six disks are placed on one spike, one on top of another, with the largest disk on the bottom, then the next largest, and so on, in order of decreasing size until the smallest disk, which is on top. (See Fig. 6-5.) You are permitted to move only one disk at a time from one spike to another spike, with the restriction that a larger disk must never be moved on top of a smaller disk. The goal is to transfer all six disks to one of the other two spikes (without ever permitting a larger disk to rest on top of a smaller disk).

I believe I was once told some relatively routine mechanical procedure for solving this problem, but I do not remember it, since I was not given a proof that indicated why it worked. However, a beautiful repeated hierarchical (recursive) use of the subgoal method provides a solution to this problem that dramatically illustrates the power of the subgoal method. Stop reading and try to solve the problem.

Given state

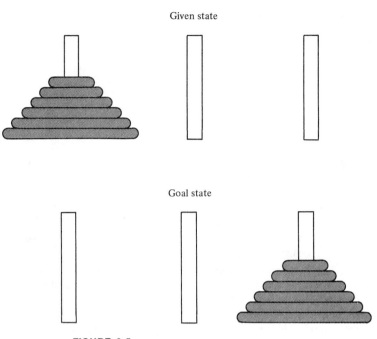

Goal state

FIGURE 6-5
The given and goal states for the Tower of Hanoi
(disk transfer) problem.

Apropos the problem-solving method (discussed in Chapters 3 and 10) of complete representation (naming) of all the concepts in a problem, the first step in solving this problem might be to give names (in this case numbers) to the disks in a manner that easily represents the one way in which they differ from each other, namely, in diameter. So let us number the disks 1 to 6 from smallest to largest. In addition, let us label the problem of transferring six disks from one spike to another a *six-problem*, which implicitly recognizes this problem as a particular case of a larger class of disk-transfer problems (five-problems, seven-problems, and so on). For convenience in verbal descriptions, let us also label the spikes *A*, *B*, and *C*. This representation of the problem is shown in Fig. 6-6. Now stop reading and try again to solve the problem, if you failed before.

Having represented the problem in this way, I think that it is reasonably likely that one would think of the following elegant way to divide the problem into subgoals. Solving a six-problem from *A* onto *C* is equivalent to solving a five-problem from *A* onto *B*, moving the six-

Given state for 6-problem   Goal state for 6-problem

```
  1                                              1
  2                                              2
  3                                              3
  4                                              4
  5                                              5
  6                                              6
 ─  ─  ─                                        ─  ─  ─
 A  B  C                                        A  B  C
```

FIGURE 6-6
The given and goal states for the Tower of Hanoi (disk transfer) problem,
with numerical representation.

disk from *A* to *C*, and solving a five-problem from *B* onto *C*. The five-
problems are equivalent to a four-problem, a move of the five-disk,
and another four-problem. In turn, the four-problems can be subgoaled
into two three-problems and a move, and so on. Thus, the entire problem can be solved by a recursive use of the subgoal method. To actually
implement the method, in this case, you must have the ability to remember what level of subgoal you are currently working on, but this
problem can easily be solved by making some notes on a piece of paper.
In any event, it is clear that this method solves the problem and, in
addition, this subgoal method gives an excellent insight into the structure of the Tower of Hanoi problem.

Story-algebra problems frequently illustrate the usefulness of the
subgoal method. Consider the following simple problem:

> Each day, Abe either walks to work and rides his bicycle home or rides
> his bicycle to work and walks home. Either way, the round trip takes one
> hour. If he were to ride both ways, it would take 30 minutes. How long
> would a round trip take, if Abe walked both ways?

Stop reading and try to solve the problem by defining some simple
subgoals.

The first subgoal you might define is to determine how long it takes
Abe to ride one way. You can then determine, as a second subgoal,
how long it takes to walk one way. Then it is trivially easy to determine
how long it takes to walk both ways and to solve the problem. You note
that the time to ride both ways is 30 minutes, from which it is obvious
that the one-way riding trip requires 15 minutes. If these 15 minutes
are subtracted from the one-hour round trip for walking plus riding,
45 minutes remain for the one-way walking trip. Doubling this yields
a round-trip walking time of 90 minutes, which is the solution to
the problem.

Most story-algebra problems are amenable to the subgoal approach. Instead of working directly to determine the value of the unknown quantity, you set subgoals of determining various other unknown quantities that are related to the goal quantity by some known relation. When all the unknown quantities except the goal quantity have been determined in the known relation, you can then use the known relation to solve for the goal quantity. You must also be able to represent the elements expressed in the story problem in algebraic (equation) form. However, after skill in algebraic representation, skill in defining subgoals is probably the next most important element in the solution of story-algebra problems.

For another simple example of the subgoal method applied to a simple story-algebra problem, consider the following:

> Ingrid brings a quantity of hats to sell at the Saturday market. In the morning, she sells her hats for $3 each, grossing $18. In the afternoon, she reduces her price to $2 each and sells twice as many. What was Ingrid's gross income for the day from the sale of hats?

Stop reading and solve the problem by defining a simple sequence of subgoals.

The first subgoal is to determine how many hats are sold in the morning. From this, it is trivially easy to determine the number of hats sold in the afternoon, which is the second subgoal, and then the gross income for the day.

The specific solution to the problem is as follows: If Ingrid grossed $18 in the morning by selling hats at $3 per hat, she evidently sold 6 hats. This implies that she sold 12 hats in the afternoon. Therefore, Ingrid grossed $6 \times \$3 + 12 \times \$2$, or $42 for the day.

The subgoal method is also frequently useful in the solution of geometry problems, as in the following example:

> Given the parallelogram $ABCD$ illustrated in Fig. 6-7, prove that the perpendiculars AE and CF drawn to the diagonal BD are equal.

Stop reading and try to solve the problem by defining a relevant subgoal.

One very common way to prove two lines are equal is to prove that they are corresponding parts of congruent triangles. In the present case, this could mean either proving that triangle ABE was congruent to triangle CDF or proving the triangle AED was congruent to triangle CFB. These two alternative subgoals appear to be equivalent, and therefore we may arbitrarily choose to work on the subgoal of proving

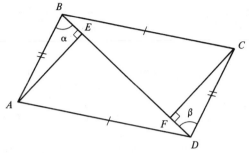

FIGURE 6-7
Given $AB = CD$, $BC = AD$, AE ⊥ BD, and CF ⊥ BD, prove that $AE = CF$.

triangle ABE congruent to triangle CDF. Stop reading and attempt to solve the problem, possibly by defining a further subgoal.

To prove triangle ABE congruent to triangle CDF, it is helpful to define a prior subgoal of proving that triangle ABD is congruent to triangle CDB. Stop reading and attempt to solve the problem, using this sequence of two subgoals.

Triangle ABD is evidently congruent to triangle CDB, since the three corresponding sides of each are equal. From this, we can conclude that angle α equals angle β in Fig. 6-7. From this, we can conclude that triangle ABE is congruent to triangle CDF, since both are right triangles and there are corresponding angles (besides the right angles) that are equal and the hypotenuses are equal. Now that these triangles have been proved congruent, side AE equals side CF by corresponding parts of congruent triangles, and the problem is solved.

Another geometry problem that is quickly solved by use of the subgoal method is the following:

> Given the circle illustrated in Fig. 6-8, proceed from the circumference along a diameter of the circle for an arbitrary unknown distance, to point A, then turn perpendicular to the radius and draw a line connecting the radius to the circumference, point B. Then erect another perpendicular at B until it intersects, at point C, the diameter perpendicular to the original diameter. The diameter of the circle is 100 feet. Determine the length of the line AC.

Stop reading and try to solve the problem by defining relevant subgoals.

At first, this problem may seem extremely difficult to solve, since very little numerical information is given in the problem. However, a reasonable subgoal for determining the length of the line AC is to determine some triangle to which ABC is congruent, where the length of one or more of the sides of the second triangle is known. Alternatively, you could attempt to determine some triangle congruent to the triangle *AOC*. Using this subgoal, attempt to solve the problem.

Of course, triangle *ABC* is congruent to triangle *AOC*; but this knowledge is not much help, since you do not know the lengths of any of the lines in either of these triangles. There are no other triangles drawn explicitly in Fig. 6-8. Thus, you will have to draw additional lines in order to define new triangles that may be congruent to the triangles already given in Fig. 6-8. Clearly, you should draw lines that result in triangles with one or more known length sides. The only known lengths are the diameters and radii of the circle. Thus, the constructed triangle should evidently include a diameter or radius. Given this line of reasoning, sooner or later you should hit upon the idea of drawing the radius *OB* to define the triangles *BOA* and *BOC*, both of which are congruent to the original triangles *ABC* and *AOC* (easily proved). From this, we conclude that line *AC* equals line *OB*, by corresponding parts of congruent triangles, and, since line *OB* is a radius of a circle, we know that line *AC* equals the radius of the circle, which is 50 feet.

The common practice in mathematics of conjecturing and proving one or more lemmas as subgoals on the way to proving some major theorem is a good example of the use of the subgoal method. The skillful definition of lemmas (subgoals) to aid in proving a difficult theorem depends on having simple and complete representations of the relevant mathematical concepts and very good evaluation functions based on such elegant representations and experience in theorem proving. Some of this ability to represent concepts elegantly and define good evaluation functions can be gained by studying general problem-solving methods and applying them to problems requiring no specialized knowledge. However, you cannot expect to be able to prove difficult theorems in some area of mathematics without extensive

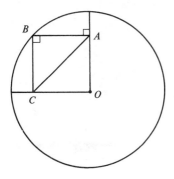

FIGURE 6-8
Given the circle illustrated above with a diameter of 100 feet, determine the length of the line *AC*.

studying of the concepts, assumptions, operations, and so on, in that area. To solve problems in a specialized area of knowledge, one must have learned certain elegant ways of representing concepts in that area. This knowledge is often required in order to define good evaluation functions for use in hill climbing and the subgoal method.

Another mathematical proof technique that is an ingenious use of the subgoal method is *mathematical induction*. Mathematical induction can be used to prove theorems that in some way involve natural numbers (positive integers). Let the goal expression you are trying to prove be represented by $E(n)$, where n stands for any natural number ($n \geq 1$). The problem of proving $E(n)$ true for any natural number, n, can be divided into two subproblems: first, proving $E(n)$ true for $n = 1$, and, second, proving that if $E(n - 1)$ is true, then $E(n)$ is true.

For example, consider a mathematical induction proof of the theorem that the sum of the first n positive integers, $\sum_{i=1}^{n} i$, is $n(n + 1)/2$:

(1) $\sum_{i=1}^{1} i = 1 = 1 \cdot \frac{2}{2}$, so true for $n = 1$.

(2) Assume true for $(n - 1)$: $\sum_{i=1}^{n-1} i = \dfrac{(n - 1)\, n}{2}$

(3) Add n to both sides: $\sum_{i=1}^{n} i = \dfrac{(n - 1)\, n}{2} + n$

(4) Put over common denominator: $\sum_{i=1}^{n} i = \dfrac{(n - 1)\, n + 2n}{2}$

(5) Factor: $\sum_{i=1}^{n} i = \dfrac{n(n - 1 + 2)}{2}$

(6) Simplify: $\sum_{i=1}^{n} i = \dfrac{n(n + 1)}{2}$ $\qquad$ Q.E.D.

Step (1) establishes that the theorem is true for $n = 1$, and steps (2) through (6) establish that if the theorem is true for $n - 1$, it is true for n.

7

Contradiction

As mentioned in Chapter 3, amateur problem solvers often do not pay enough attention to the goal or the set of possible goals as part of the information in a problem. They apply operations to the givens in an attempt to get to the goal, but they frequently do not consider applying operations to the possible goals in order to get to the givens or to meet the givens halfway.

In Chapter 3 we were concerned with inferences about the goal that could be made primarily from the partial information the solver already possessed about the goal, but also from given information or from information about both givens and the goal. Here we are also concerned with inferences that can be made from the goal in conjunction with the givens. However, the purpose of the types of inferences I will discuss here is quite different from the purpose of those in Chapter 3, where the purpose was to clearly specify the parts of the goal, so that we could more easily see exactly what was to be derived from the givens. By contrast, the purpose of the types of inferences to be discussed now is to derive an inference that contradicts some piece of given information.

Deriving a *contradiction* proves that the goal could not possibly be obtained from the givens, since it is inconsistent with the givens. This method of contradiction is appropriate for several types of problems in which you must decide which of two or more goals could be derived from the givens. The method of contradiction only tells you which goals *cannot* be derived from the givens. However, in some problems, the ability to decide whether a possible goal does or does not contradict the givens may be all that is required to solve the problem.

Problems for which the method of contradiction is appropriate include those where you must only determine whether a goal is inconsistent with the given information, not necessarily whether it could be derived from the given information using some particular set of operations. The method is also appropriate for problems that guarantee that exactly one of several alternative goals can be derived from the given information. Here if all alternatives but one can be ruled out, then that one must be derivable from the givens, and the method of contradiction constitutes a sufficient proof of it.

Many problems make the guarantee that one out of several alternative goals can be derived from the givens. In this chapter, I will discuss in four sections the problems that illustrate how the method of contradiction can be applied. The four sections are as follows.

Indirect Proof   The first section will be concerned with problems with only two or three alternative goals. It will focus on the method of indirect proof in mathematics, where the two alternative goals are usually that some statement is either true or false. We are not interested in whether we can always say a statement is either true or false but rather in the use of the method of contradiction in those cases where it is assumed that only these two alternatives can hold. In such cases, if a person can show that one of the two alternatives leads to a contradiction, then the other alternative has been proved.

Multiple Choice — Small Search Space   The second section will be concerned with the method of contradiction in problems involving a small (from two to 10) set of alternative goals that are mutually inconsistent (only one of the goals can be derived from the givens). In problems involving a small set of alternative goals, it is feasible to systematically apply the method of contradiction to every alternative goal. Examples of such problems include multiple-choice examination problems and certain logic problems.

Classificatory Contradiction — Large Search Space   The third section will be concerned with the use of the method of contradiction in problems in which there is a large, but discrete and finite, population of alternative goals. In these problems, it is generally not feasible to systematically search every alternative. It is necessary to devise some more efficient search procedure that contradicts large classes of alternative goals simultaneously. Problems in this category include the coin-weighing problem discussed earlier, many concept-attainment problems, and letter-arithmetic problems.

Classificatory Contradiction—Infinite Search Space The fourth section will be concerned with the use of the method of contradiction in problems involving infinite (often continuous) populations of goals. In these problems, it is clearly impossible to contradict each goal individually; the solver must contradict infinitely large classes on the basis of some common property. An example of this case is provided by the half-interval search technique in the numerical solution for roots of equations.

INDIRECT PROOF

The method of indirect proof in mathematics is an extremely important example of the problem-solving method of contradiction. To prove that a statement follows from certain givens, the method of indirect proof is to assume the contrary is true and show that the contrary statement, in combination with the givens, results in a contradiction. Therefore, since the contrary statement is false, the original statement must be true. You will note that, in this case, for the method of indirect proof to be valid there must be only two possible alternatives: either the goal statement is true (can be derived from the givens) or it is false (the contradiction of the statement can be derived from the givens, but the original statement cannot). For the method of contradiction to be valid, the statement must be either true or false. The truth value of the statement cannot be undecidable. In addition, the set of given statements must themselves be free of internal contradiction; otherwise, contradictions could be derived from a possible goal in combination with the givens, not because of a contradiction between the goal and the givens, but because of a contradiction within the givens. However, the beginning student need not be very concerned with these limitations on the use of the method of contradiction. By and large, whenever it appears reasonable to use the method of contradiction, it is valid to use it.

There are, of course, innumerable examples of the use of indirect proof in every area of mathematics. Here is one example:

Given that you have already proved the theorem that all squares of non-zero integers are positive, prove that equation $x^2 + 1 = 0$ has no integer solution.

Stop reading and attempt to prove the theorem, using the method of contradiction (indirect proof).

The first step in applying the method of contradiction to this problem is to assume the contradiction of the theorem — namely, that x has an integer solution, $x = c$, where c is an integer. If you have so far not solved the problem, stop reading and try again.

Given that $x^2 + 1 = 0$, subtract 1 from both sides of the equation to get $x^2 = -1$. Now substitute c for x, getting $c^2 = -1$. This result is a contradiction to the already proved theorem that the square of any integer must be positive.

A famous proof of the existence of irrational numbers also uses the method of contradiction. Rational numbers are numbers expressible by simple fractions, m/n, of integers m and n, where n is nonzero. To show the existence of irrational numbers, we need to show that there exists at least one such number — for example, $\sqrt{2}$.

> Given an isosceles right triangle with sides of unit length, the Pythagorean Theorem asserts that the length of the hypotenuse equals the square root of the sum of the squares of the lengths of the sides; namely, $c = \sqrt{1^2 + 1^2} = \sqrt{2}$. Prove that the length of the hypotenuse of this triangle — namely, $\sqrt{2}$ — is irrational.

Stop reading and try to solve the problem, using the method of contradiction.

The contradiction of the theorem is to assume that the $\sqrt{2}$ is rational and therefore can be expressed as the ratio of two integers m/n, where both m and n are integers (greater than zero). Also, when we assume $\sqrt{2}$ equals m/n, we can assume that m and n have no common factors, since these common factors could already have been canceled out. If you have not solved the problem so far, stop reading and try again.

From the above, we derive that $2n^2 = m^2$, which implies that m^2 is even. This result in turn implies that m is even ($m = 2p$, where p is an integer). If you have still not solved the problem, stop reading and try again.

If $m = 2p$, we can substitute $2p$ for m in the equation $2n^2 = m^2$, obtaining $2n^2 = (2p)^2 = 4p^2$. From this result we obtain $n^2 = 2p^2$, which implies that n^2 is even. Again, this result implies that n is even (contains a factor of two). However, we have now derived that both m and n are even (contain a common factor of 2), contradicting the hypothesis that m and n have no common factors. Thus, the contradiction is false, and $\sqrt{2}$ must be irrational.

A common feature of both these examples of indirect proofs that is characteristic of most proof problems to which the method of indirect proof is well suited is that the contradiction of the theorem permits a larger number of specific consequences to be derived from it than does the original statement of the theorem. This feature gives you a great deal more to work with by concentrating on the contradiction of the theorem than you would have by concentrating on the original statement of the theorem. It should be clear, then, why the method of contradiction is so useful in cases like this.

The method of contradiction is also used in proof problems where there are two or more incorrect alternatives to the correct theorem, each of which must be disproved by contradiction when combined with the givens. For example, consider a proof of the following theorem:

> You are given three assumptions or previously proved theorems. (1) If $c > 0$ and $a = b$, then $ac = bc$. (2) If $c > 0$ and $a > b$, then $ac > bc$. (3) The law of trichotomy obtains: for any a and b, one and only one of three alternatives holds: $a < b$, $a = b$, or $a > b$. Using these givens, prove that, if $c > 0$ and $ac < bc$, then $a < b$.

Stop reading and try to prove the theorem, using the method of contradiction.

To prove this theorem, you must test two incorrect alternatives to show that they result in contradictions — namely, $a > b$ and $a = b$. If you have not already proved the theorem, stop reading and try again.

First, let us assume $a > b$. If $a > b$, then with $c > 0$, we know that $ac > bc$. But this result is a contradiction to the given information that $ac < bc$, by the law of trichotomy. Similarly, to rule out the alternative that $a = b$, we derive from $a = b$ and $c > 0$ that $ac = bc$, which contradicts the given information that $ac < bc$, by the law of trichotomy. Therefore, the only remaining possibility, by the law of trichotomy, is that $a < b$, which was to be proved. Note that we had to rule out *two* alternatives before we could conclude the theorem proved, although in this case the method of contradicting each alternative was extremely similar.

The method of indirect proof shows up in an enormous variety of problems. For example, recall that one essential part of the solution to the notched-checkerboard problem in Chapter 3 was to assume that there was a method of covering 62 squares with 31 dominoes. These 31 dominoes must cover 31 black squares and 31 white squares. From the given information, we can derive that removing the two diagonally opposite squares of the checkerboard will produce 32 squares of one color and 30 squares of the other color, resulting in a contradiction. Thus, there is no method of covering the 62 remaining squares with 31 dominoes.

Similarly, in the method used in Chapter 3 to establish the minimum number of cuts needed to solve various cube-cutting problems, we implicitly ruled out a smaller number of cuts by contradiction with the implicit given information that not more than one face of a subcube could be cut at a time.

The method of indirect proof is often useful in plane geometry proof problems. Indeed, high school plane geometry is usually the first opportunity for most students to become acquainted with the method of indirect proof. One reason why most people are so suspicious of the method of indirect proof when they first encounter it is that they encounter it so late. I suspect they unconsciously feel that so basic a method of proof should have been explained to them much earlier in their lives, as indeed it should have been. Be that as it may, plane geometry proof problems often demonstrate the method of indirect proof, and the following is a particularly simple example:

> Given the assumption that two distinct points determine one and only one straight line, prove that two lines can intersect at no more than one point.

Stop reading and try to prove this theorem, using the method of contradiction.

First, assume the contrary—namely, that there exist two lines that intersect in at least two points, A and B. If you did not solve the problem thus far, stop reading and try again.

Since the two straight lines intersect in points A and B, there are two distinct straight lines passing through the points A and B. However, this is contrary to the assumption that two points determine one and only one straight line. Thus, the contrary of the theorem is contradicted, and so the theorem is proved.

Finally, consider the following plane geometry problem as an example of the use of indirect proof:

> You are given two assumptions or previously proved theorems. (1) A straight line is a 180° angle. (2) Two lines are perpendicular, if they make a 90° angle where they intersect. From these assumptions, prove that from a point on a line, only one perpendicular line can be erected.

Stop reading and try to prove this theorem, using the method of contradiction.

To prove this geometric theorem it is useful, as it almost invariably is in plane geometry problems, to construct a figure. Consider Fig. 7-1. To prove that at most one line can be perpendicular to another line at a given point, assume the contrary—namely, that at least two lines can be drawn perpendicular to a given line through a given point. In

Fig. 7-1 this assumption is represented by the two lines drawn through point A and represented as being perpendicular to line C. If you have not yet proved the theorem, stop reading and try again. According to the hypothesis that the two perpendiculars are distinct, there is some angle α between them, $\alpha > 0$. Each of the perpendiculars forms a 90° angle with line C. Thus, the straight line C equals 90° $+ \alpha + 90° > 180°$, which is contrary to the assumption that a straight line is a 180° angle.

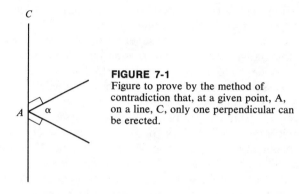

FIGURE 7-1
Figure to prove by the method of contradiction that, at a given point, A, on a line, C, only one perpendicular can be erected.

MULTIPLE CHOICE—SMALL SEARCH SPACE

Besides being useful for proving theorems, the method of contradiction is useful in the solution of a wide variety of other problems where there are usually more than two alternatives. Whenever you are guaranteed that exactly one of a small set of alternative goals is consistent with (or follows from) the given information, it is possible to determine which goal, by systematically examining each and deriving a contradiction from all but one of them.

Problems given on tests with multiple-choice answers have so few choices (five or less) that contradiction is frequently the ideal solution method. Simply take each alternative answer in turn and determine whether it is consistent with the given information. That is, combine each possible answer with the given information to attempt to derive a contradiction. If you can derive contradictions for all the answers except one, then that remaining answer is correct. For example, consider the following potential exam problem:

The solution of $\sqrt{7x - 3} + \sqrt{x - 1} = 2$ is: (A) $x = 3$, (B) $x = \frac{3}{7}$, (C) $x = 2$, (D) $x = 1$, (E) $x = 0$.

Stop reading and try to solve this problem, using the method of contradiction.

The slow way to solve the problem is to perform various operations on both sides of the equation (adding, subtracting, multiplying, dividing, squaring both sides of the equation). The fast way is to substitute each of the alternative values of x into the equation and see which ones work. In this case, only $x = 1$ is consistent with the equation, so (D) is the answer.

Solving for the values of one or more variables that satisfy one or more equations is a primary example of problems where often the givens and the goal should be combined. In these problems, there may be several values of a variable, or several sets of values of the several variables, that satisfy the equation or equations. You are being asked to determine one or more such solutions that satisfy the equations, which is in essence saying that consistency of the givens and goal statements is all that is demanded.

When you encounter a problem like this on a multiple-choice test, with a small number of choices for solutions of the equations, then the ideal problem-solving method is contradiction. You simply try each of the alternative solutions in turn to see if it satisfies the equations—that is, gives an answer like $a = a$ for all equations—when the set of values of variables is substituted into the equations. All of the incorrect answers will produce contradictions such as $5 = 3$. For example, consider the following problem:

Which of the following is a solution of the cubic equation, $x^3 + 4x^2 - 7x - 10 = 0$? x equals: (A) -2, (B) -5, (C) 4, (D) 3, (E) none of these.

Stop reading and try to solve this problem, using the method of contradiction.

In this instance, you can factor the cubic into three linear factors, yielding three real roots. However, a much faster way to solve the problem is to check each of the first four specific alternative answers for consistency with the cubic equation.

Since one of the answers—namely, $x = -5$—is consistent with the cubic equation, and all the rest are not, we know that (B) is the correct answer. Since all you are asked is whether a particular alternative goal is consistent with the given information (the cubic equation), determining one answer that is consistent is sufficient to solve the problem. You need not actually derive contradictions in the case of alternatives (C) and (D), except as a check that your determination of consistency in the case of alternative (B) was correct. If alternative

(E) was an expression such as "several of these," it would be necessary to derive contradictions to all but one of the first four alternative answers in order to rule out this alternative.

Contradiction is useful in examinations on a very wide variety of problems. For example, consider the following:

> The base of our number system is 10. If the base were changed to four you would count as follows: 1, 2, 3, 10, 11, 12, 13, 20, 21, 22, 23, 30, and so on. The 22nd number in the base-four system is: (A) 22, (B) 37, (C) 64, (D) 104, (E) 112.

Stop reading and try to answer this question, using the method of contradiction.

In a number system with base of four, the digits 4, 5, 6, 7, 8, and 9 cannot appear. Thus, alternative answers (B), (C), and (D) could not possibly be correct. In addition, the number 22 has already been used prior to achieving the 22nd number in the base-four system. Thus, the only possible answer of the five that could be correct is answer (E), which is 112.

Some problems are necessarily solved by contradiction. For example, consider the following problem:

> The formula expressing the relationship between x and y in the table is:

x	1	2	3	4	5
y	2	3	6	11	18

> (A) $y = 2x + 1$, (B) $y = -x^3 + 2x^2 + 1$, (C) $y = x^4 - 2x^3 + 3x^2 - x + 1$, (D) $y = x^2 - 2x + 3$, (E) $y = x^2 + 1$.

Stop reading and try to solve this problem, using the method of contradiction.

The correct answer is alternative (D), and there is really no other way to determine that (D) is the correct answer except by using the method of contradiction. Infinitely many different functions are consistent with any finite number of points, and infinitely many different functions are inconsistent with any finite number of points. All we can do is to determine which functions are consistent and which are inconsistent by checking for contradictions between the proposed functions (goals) and the given information about points on the function.

No one would object to your using the method of contradiction to solve such problems in an examination situation and, of course, to using it in the case of indirect proof. However, I have occasionally heard objections to its being used in some of the above examples of multiple-choice problems, where there are direct methods for determining the goal from the given information. Some teachers have even protested that for students to use the answers and look for contradictions is mildly immoral, that it "educates people to be test takers."

It is true that such students are not demonstrating their knowledge of the direct algorithmic methods for obtaining the goal from the given information. However, I think we must face the fact that these types of test questions simply do not adequately assess a student's knowledge of algorithmic methods, because the method of contradiction can be used in place of the algorithms. Some students are inevitably going to use the method of contradiction whether anybody tells them about it or not, and this only introduces an extra source of noise in the relative assessment of understanding of different students. Test questions must be made fool-proof; it will not do to ask students to be fools. Furthermore, there are many times when a teacher wants to assess students' knowledge of certain specific mathematical concepts with questions that can be answered using the method of contradiction, and it is either not possible or the teacher does not care to assess their understanding of any algorithmic method for generating the solution.

Another class of problems that involves a search among a small population of alternative goals are the recreational logic problems that make up such a large part of problem books. In many of these problems it is difficult to make inferences from the givens to the goal, but it is generally quite easy to test any given assumption about the goal for consistency with the given information. Since the number of alternative goals is often quite small, the method of contradiction is ideally suited to the solution of such problems. Some of the most interesting recreational logic problems that are the most difficult to solve by inferences from the givens to the goal are problems involving the possibility that some of the given information is false. These are the famous liar and truth-teller (truar) problems. We discussed one such problem in Chapter 3 in connection with the need for having a clear understanding of the goal; let us consider it again here from the standpoint of the method of contradiction:

> The country of Marr is inhabited by two types of people, liars and truars (truth tellers). Liars always lie and truars always tell the truth. As the newly appointed United States ambassador to Marr, you have been invited to a local cocktail party. While consuming some of the native spirits, you are engaged in conversation with three of Marr's most prominent

citizens: Joan Landill, Shawn Farrar, and Peter Gant. At one point in the conversation Joan remarks that Shawn and Peter are both liars. Shawn vehemently denies that he is a liar, but Peter replies that Shawn is indeed a liar. From this information, can you determine how many of the three are liars and how many are truars?

Stop reading and try to solve this problem, using the method of contradiction. You will recall from Chapter 3 that all we need to determine is how many of the three are liars—namely, whether there are zero, one, two, or three liars among the three people. However, in order to determine this number, it is useful to consider all eight possibilities for the liar-versus-truar status of Joan, Shawn, and Peter—namely, all three are liars; Joan and Shawn are truars, but Peter is a liar; Joan and Peter are liars, but Shawn is a truar; Shawn and Peter are liars, but Joan is a truar; Joan is a liar, but Shawn and Peter are truars; Shawn is a liar, and Joan and Peter are truars; and, finally, Peter is a liar, and Joan and Shawn are truars. It is easy to test the consistency of each of these eight possibilities with the given information of each of these eight possibilities. For example, all three cannot be truars, since Joan would not then say that both Shawn and Peter were liars. All three cannot be lying, since Peter would not then say that Shawn was a liar. We can also rule out each of the three possibilities in which there are one liar and two truars in the group. Of the remaining three possibilities, we can rule out the possibility that Joan is a truar and both Shawn and Peter are liars, but we cannot find a contradiction to either of the other two possibilities—namely, that Joan and Shawn are liars and Peter is a truar, or that Joan and Peter are liars and Shawn is a truar. As discussed in Chapter 3, the inability to decide between these two possibilities is of no consequence to the solution of the original problem, since all we were asked to determine was how many of the three are liars. Under either of the two possibilities that are not contradictory with the given information, there are exactly two liars and one truar, which is the answer to the problem.

Instead of examining all eight combinations of liar-and-truar status for each of the three people, it is possible to use the method of contradiction somewhat more efficiently by considering various classes of the eight alternatives. For instance, a judicious choice would be to consider the class of possibilities in which Joan is a truar. All of the four members of this class can be shown to be contradictory to the given information, since then both Shawn and Peter must be liars and Shawn would then be telling the truth—a contradiction. This classificatory use of the method of contradiction is discussed more extensively in the next section. In problems involving search through only a small set of alternatives, it is usually quickest to test each of the possibilities individually for consistency with the given information.

Another liar-truar problem that simply illustrates the usefulness of the method of contradiction is as follows:

> The Nelsons have gone out for the evening, leaving their four children with a new babysitter, Nancy Wiggens. Among the many instructions the Nelsons gave Nancy before they left was that three of their children were consistent liars and only one of them consistently told the truth, and told her which one. But in the course of receiving so much other information, Nancy forgot which child was the truar. As she was preparing dinner for the children, one of them broke a vase in the next room. Nancy rushed in and asked who broke the vase. These were the children's statements:
>
> *Betty:* Steve broke the vase.
> *Steve:* John broke it.
> *Laura:* I didn't break it.
> *John:* Steve lied when he said I broke it.
>
> Knowing that only one of these statements was true, Nancy quickly determined which child broke the vase. Who was it?

Stop reading and try to solve the problem, using the method of contradiction.

There are two possible approaches to this problem. First, we might try to test each of the four possibilities for who broke the vase. This approach appears to be the most direct way to the goal; however, it will not work until we first determine which of the four is telling the truth and which three are lying. When the liar-versus-truar status of the four children has been determined, it is trivial to determine who broke the vase. Thus, to successfully apply the method of contradiction to the problem, we should test the four possibilities in regard to which of the children is a truar. If you did not solve the problem before, stop reading and try again, using this indirect application of the method of contradiction.

Betty cannot be the truar, since then both Betty and Laura would be telling the truth, contrary to the information that only one can be telling the truth. For the same reason, Steve could not be telling the truth, since then both Steve and Laura would be telling the truth. Laura cannot be telling the truth because then, if John is lying, Steve is telling the truth, contrary to the information that only one child can be telling the truth. The only possibility that is consistent with the given information is that John is telling the truth and Betty, Steve, and Laura are lying. Given this, it is trivial to determine that Laura must be the one who broke the vase.

One of my all-time favorite recreational logic problems is the famous *Smith, Jones, and Robinson problem*:

> Smith, Jones, and Robinson are the brakeman, fireman, and engineer of a train, not necessarily respectively. Today only three passengers are riding this train, and, by an extraordinary coincidence, their last names are the same as the last names of the brakeman, fireman, and engineer. To distinguish the passengers from the trainmen, let us refer to the passengers with the title Mr. — Mr. Smith, Mr. Jones, and Mr. Robinson. Here is some other relevant information:
>
> (A) Mr. Robinson lives in Detroit.
> (B) The brakeman lives halfway between Chicago and Detroit.
> (C) The passenger who lives in Chicago has the same name as the brakeman.
> (D) The brakeman's next-door neighbor, one of the passengers, earns exactly three times as much as the brakeman.
> (E) Mr. Jones earns exactly $2,000 a year (and collects a lot of food stamps and welfare payments).
> (F) Smith beat the fireman at billiards.
> Who is the engineer?

Stop reading and try to solve the problem, using the method of contradiction.

The most direct application of the method of contradiction to this problem would be to test the three possibilities for the name of the engineer — Smith, Jones, or Robinson — against the given information. As in the previous problem, this most direct approach is not the best, since none of the six statements of information includes any reference to the engineer. Thus, it is obvious that if we are to determine the name of the engineer, we must consider some more indirect approach, which first involves determining who might be the brakeman or the fireman, who might live next door to whom, who might live in what city, and so on. If you did not solve the problem thus far, stop reading and try again.

A minimal expansion of the search space of alternatives, using the method of contradiction, is to consider each of the six possibilities for the assignment of names to the brakeman, fireman, and engineer, as illustrated in the following table (S = Smith, J = Jones, R = Robinson):

Person	Hypotheses					
	1	2	3	4	5	6
Brakeman	S	S	J	J	R	R
Fireman	J	R	S	R	S	J
Engineer	R	J	R	S	J	S

Now we examine the six pieces of information to determine which of these six possibilities produces a contradiction and therefore can be eliminated from consideration. In the first place, hypotheses 3 and 5 can be eliminated, because condition (F) says that Smith beat the fireman at billiards; assuming a person cannot beat himself, then, Smith cannot be the fireman. All but one of the remaining four possibilities can be eliminated by verbal reasoning, but it can be a little confusing.

A great deal of the information in the problem concerns the passengers and, in particular, where they live. Thus, it probably would be helpful to go a step further away from the direct approach to the goal and to try to test various possibilities for the assignment of passengers' names to locations. If you have not solved the problem already, stop reading and try again, using the method of contradiction as applied to the various possibilities for home addresses of the three different passengers.

There are three home addresses for the passengers – Chicago, Detroit, and halfway between Chicago and Detroit. Furthermore, one and only one passenger lives in each of the three locations. Since the given information says that Mr. Robinson lives in Detroit, there are only two remaining possibilities for the complete assignment of passengers to home addresses: either Mr. Jones lives in Chicago and Mr. Smith lives between Chicago and Detroit or else Mr. Smith lives in Chicago and Mr. Jones lives between Detroit and Chicago. Since Mr. Jones earns *exactly* $2,000, and $2,000 is not divisible by 3, and the brakeman's next-door neighbor earns exactly three times as much as the brakeman, Mr. Jones cannot live next door to the brakeman (halfway between Chicago and Detroit). Thus, Mr. Jones must live in Chicago and Mr. Smith must live halfway between Detroit and Chicago. Since Mr. Jones lives in Chicago, the brakeman is Jones by statement

(C). This result eliminates, by contradiction, alternatives 1, 2, 5, and 6 in the assignment of names to the three positions of brakeman, fireman, and engineer. Since we already ruled out alternative 3, we are left with only alternative 4, consistent with the given information. Thus, Smith is the engineer (Jones is the brakeman and Robinson the fireman).

It often facilitates work on recreational logic problems of this type to set up various tables representing what goes with what. In the present instance, there are two useful tables. First is a table such as the following, involving the assignment of the names (Smith, Jones, Robinson) to the positions (brakeman, fireman, engineer):

	Smith	Jones	Robinson
Brakeman			
Fireman			
Engineer			

In addition, it is useful to set up a table assigning passengers' names to home addresses, as follows:

	Mr. Smith	Mr. Jones	Mr. Robinson
Chicago			
Detroit			
Halfway between			

When you acquire a piece of information such as that Smith cannot be a fireman, you enter no in the box of the table appropriate to Smith being a fireman. When you know from given information that Mr. Robinson lives in Detroit, you enter yes in that box of the relevant table; you also enter no in every other box in the same row or column of the table, since there can be only one yes in each row or column of such logic tables. It is the restriction to only one yes in a row or column that permits rather powerful use of this tabular representation: whenever you have a yes in a row or column, you can fill in the rest of both the row and the column with nos; whenever you have two nos in a row or column, you can fill in a yes in the remaining position in that row or column. Tabular representation permits us to draw inferences quite mechanically from previous inferences that are recorded in the table, avoiding complicated verbal reasoning and the possibility of memory loss. The finished versions of these two tables for solution of this Smith, Jones, Robinson problem are shown in Fig. 7-2.

	Smith	Jones	Robinson
Brakeman	No	Yes	No
Fireman	No	No	Yes
Engineer	Yes	No	No

	Mr. Smith	Mr. Jones	Mr. Robinson
Chicago	No	Yes	No
Detroit	No	No	Yes
Between	Yes	No	No

FIGURE 7-2
Final tables for solution of Smith, Jones, and Robinson problem.

A final problem of a completely different kind that illustrates the usefulness of the method of contradiction is a spatial-puzzle problem that I have called the *bowling-pin reversal problem*:

> Six-year-old Heather Phillips set up the ten pins for her bowling game at the end of the hall in a manner exactly opposite to the correct configuration. Before Heather could throw the bowling ball down the hall, her father informed her that she had set up the pins in the wrong manner and that the pins should have the row of one pin in front, followed by the row of two pins, followed by the row of three, and, finally, the row of four in the back. Although Heather is given to childish reversal errors of this type when she forgets to put on her thinking cap, she is actually a budding mathematical genius. So, upon being informed of her error, Heather quickly put her thinking cap back on, ran down the hall, and, by moving just three pins, was able to reverse the configuration from the given state to the goal state, as illustrated in Fig. 7-3. How did she do it? (By the way, Heather assumed that the exact placement of the pins on the floor was not important, so long as the relative placement of the pins with respect to each other was correct. You should assume this also.)

Stop reading, put on your thinking cap (if you do not have it on already), and try to solve the problem, using the method of contradiction (not just random trial and error).

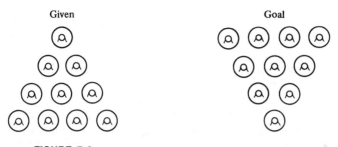

FIGURE 7-3
Given and goal states for the bowling-pin reversal problem.

To apply the method of contradiction, we need to have a well-defined set of possibilities. The smallest such set of possibilities is to ask where the row of four pins will be in the goal state with respect to its position in the given state. There are six logical possibilities for this — above the row, with one pin in the given state; the row with one pin in the given state will become the row with four pins in the goal state; the row with two pins in the given state will become the row with four pins in the goal state; the row with three pins in the given state will become the row with four pins in the goal state; the row with four pins in the given state will remain the row with four pins in the goal state; or the row of four pins in the goal state will be below the row with four pins in the given state. If you have not solved the problem thus far, stop reading and try again, using the method of contradiction to eliminate all but one of the six possibilities.

Clearly, if the row of four pins were either above the row of one pin or below the row of four pins in the given state, we would have to move more than three pins in order to achieve this aspect of the goal state alone. Thus, we have contradicted these two possibilities. To make the row of one pin in the given state the row of four pins in the goal state would require a minimum of three moves to achieve that subgoal alone, plus the row of two pins would then have to become the row of three pins, producing already more than three moves. This arrangement cannot be the desired solution. If the row of four pins in the given state is to remain the row of four pins in the goal state, all six pins above it would have to be moved, contradicting the requirement of the proposed solution. Finally, if the row of three pins were to become the row of four pins in the goal state, all three pins above the row of three would have to be removed, plus one of the pins in the row of four would have to be moved, contradicting the restriction to only three moves. This result leaves only the possibility of making the row of two pins in the given state into the row of four pins in the goal state, producing the solution in a fairly direct manner: the two extreme pins in the row of four are moved to the two extreme positions in the row of two, and the top pin in the given state is moved to the middle of the row, below all the other pins, achieving the goal state.

Probably the most common way of solving this problem is not to use the method of contradiction but rather to look for a subset of seven pins in the given state in identical positions to seven pins in the goal state. In general, if you did not know the minimum number of moves to transform the given state into the goal state in a problem of this type, you might look for the maximum subset of entities in the given state that were in identical positions *relative to one another* to positions in the goal state. Implementation of this method is largely a perceptual method of scanning the given and goal states looking for matches of (usually compact) subsets.

CLASSIFICATORY CONTRADICTION – LARGE SEARCH SPACE

In discussing the method of contradiction above, we were able to conceptualize the problems so that there was a relatively small population of alternative goals to decide among by the method. In the problems here, however, the number of alternative specific goals is so large that contradicting them one at a time would be impractical. In such cases, we must use some efficient search strategy for contradicting large subgroups of alternative goals at a time. To implement this more efficient search, some explicit or implicit classification must be imposed on the alternative goals, and classes of goals must be contradicted on the basis of common properties possessed by all of them. In addition, there is often some natural ordering for the contradiction of different classes of goals such that it is easiest to rule out a particular class in the beginning, some other class next, another class next, and so on. The ruling out of earlier classes of goals provides the additional information necessary for contradicting subsequent classes of goals. Attempts to rule out classes of goals in orders other than the natural or easiest ordering will usually be extremely difficult or impossible.

The coin-weighing problems discussed in Chapters 3 and 5 are examples of the method of classificatory contradiction. Recall that in the simplest of these problems, you must determine which of n coins is the heavy coin, using a beam balance. Weighing one group of coins against another provides information that contradicts a large class of possibilities with respect to which coin is the heavy coin. Whether you consider these problems to exemplify the contradiction of one set of alternatives or the implication of the complementary set of alternatives is obviously completely arbitrary.

A class of problems somewhat similar to the coin-weighing problems in the need for classificatory contradiction of alternative goals are the concept-attainment problems, of which the following is one example:

> You are given a set of six-place numbers (for example, 792,674, which is to be read as 7 in place 1, 9 in place 2, and so on), some of which are examples of the concept and some of which are not. Concepts are either *simple concepts* of the form "concept is d in place p" (that is, a particular digit d in a particular place p) or *conjunctive concepts* of the form "concept is d_1 in place p_1 and . . . and d_i in place p_i" (that is, a conjunction of digits in particular places). If the concept were 9 in place 2 and 7 in place 5, then 792,674 would be an example of the concept, because it meets both necessary conditions. On the other hand, 722,674 would not be an example of the concept, because it lacks one of the necessary conditions: it does not have a 9 in place 2. Now, determine the conjunctive concept that is implied by the following information concerning some six-place numbers that are known to be examples and nonexamples of the concept:
>
> > 107,254 is an example of the concept.
> > 157,254 is an example of the concept.
> > 937,254 is an example of the concept.
> > 867,184 is an example of the concept.
> > 295,684 is not an example of the concept.
> > 367,497 is not an example of the concept.

Stop reading and try to solve the problem, making classificatory use of the method of contradiction.

From the first piece of information that 107,254 is an example of the concept, we can determine that the concept will include some combination of the following six restrictions: 1 in place 1, 0 in place 2, 7 in place 3, 2 in place 4, 5 in place 5, and 4 in place 6. Rather than test all 63 different subsets of combinations of from one to six of these restrictions, it is much more efficient to test each of the six restrictions individually—namely, test whether a concept must involve the restriction of 1 in place 1, and so on. This procedure in essence amounts to testing the class of all concepts that involve the restriction 1 in place 1. Stop reading and solve the problem, if you have not done so already.

Clearly, the set of concepts that involve the restriction 1 in place 1 is contradicted, because some of the examples do not have the 1 in place 1. Proceeding in the same manner, we can rule out all concepts except those that require 7 in place 3 or 4 in place 6. What information tells you that both of these restrictions are necessary in order for a six-place number to be an example of the concept? This information comes from the two nonexamples of the concept, each of which illustrates that either of the restrictions in isolation is not sufficient to make a six-place number an example of the concept. Both are required.

Classificatory contradiction in concept-attainment problems is equivalent to deriving some rules of inference as to which dimensions or places of the examples of the concept are *relevant* to the concept and which are *irrelevant*. For simple and conjunctive concepts of this type, there are two very simple rules: (a) If two examples of the concept differ in the values or digits they have on one or more dimensions (places), all of these dimensions are irrelevant to the concept (that is, not involved in the necessary conditions specified by the concept). (b) If an example of the concept and a nonexample differ on one and only one possibly relevant dimension, then that dimension is relevant (and the value of that dimension in the example is the necessary value). Having derived these rules of inference, we can now solve all simple and conjunctive concept-attainment problems in a very straightforward manner. Classificatory contradiction is essentially equivalent to this inference method.

Letter-arithmetic problems, such as that below, nicely illustrate how classificatory contradiction frequently can be combined with drawing inferences to provide a solution. I got the following problem from Bartlett (1958) and Simon and Newell (1971), who have studied how people solve this problem:

$$\begin{array}{r} DONALD \\ + GERALD \\ \hline ROBERT \end{array}$$

This problem is to be treated as an exercise in simple addition. All that is known is the following: (1) $D = 5$, (2) every number from 0 to 9 has its corresponding letter, (3) each letter must be assigned a number different from that given for any other letter. The goal is to find a number for each letter, stating the steps of the process and their order.

Stop reading and try to solve the problem.

Here you should use classificatory contradiction, which was suggested as an optimal method for the concept-attainment problems. You should test hypotheses concerning the values of each letter, which is, in essence, the testing of classes of hypotheses about how all the letters are assigned to different numbers. That is, in testing the hypothesis that $N = 3$, you are, in essence, testing the entire set of possible solutions to the problem in which $N = 3$ and the other letters equal various other digits. If you did not solve the problem, stop reading and try again.

By knowing that $D = 5$, we can infer that $T = 0$. Thus, from the $O + E = O$ column, we know that E must equal 9, there having been a carry of 1 from the previous $N + R = B$ column. Since there is a carry from $D + D = T$ to the next column, we know that $L + L + 1 = R$ must be an odd number. From the $D + G = R$ column, we know that R is a number greater than 5. Thus, R could only be the number 7, since we have ruled out every other possible hypothesis. Now, since $E = 9$ and $A + A$ must be an even number, we know there had to be a carry from the $L + L = R$ column to the $A + A = E$ column. Therefore, either A could be 4, so that $4 + 4 + 1 = 9$, or A could be 9, so that $9 + 9 + 1 = 19$, that is, 9 plus a carry. However, 9 is already used. Thus, we know that A can only be 4. As Simon and Newell (1971) point out, you can proceed to determine a unique number for each letter except N, B, and O. For these numbers, you must actually try out the various combinations of remaining digits, 6, 3, and 2, assigning them to the three letters in each of the six possible ways and testing whether each assignment is consistent with the information given in the problem. If it is not, you must try a new assignment of the three remaining letters to the three remaining numbers, until an assignment is found that works.

Clearly, this last stage is contradiction pure and simple. However, in many of the preceding inferences you used the method of contradiction: you determined which letter worked by ruling out all possible alternative assignments of digits to that letter. If you have not done so, you would not have known for sure that the digit that seemed to work was the only digit that would work when assigned to that letter. Notice, however, that an efficient solution to the problem requires classificatory contradiction. In this problem, then, you must determine as nearly as possible what number to assign to a given letter, independently of testing hypotheses about the numbers to be assigned to other letters. This procedure dictates that the values of different letters must be specified in a certain order because only in certain orders is it possible to determine a unique number to be assigned to each letter.

As illustrated in letter-arithmetic problems and concept-attainment problems, classificatory contradiction is somewhat analogous to the problem-solving method of defining subgoals in those problems requiring the construction of a long sequence of actions in order to achieve the goal. In contradiction problems, the difficulty is not in the long sequence of operations but in the large class of possible hypotheses. But either way, you have in essence a large set of alternatives to search through. To the extent that you can reduce that search space by considering large classes of alternatives at one time, it is advantageous to do so.

Another problem that exemplifies the combined use of inferences and contradiction to reduce a large number of hypotheses to a small number is the *integer-path-addition problem*:

> Put the digits 1, 2, . . . , 9 into a 3 × 3 matrix, one digit into each cell, as shown in Fig. 7-4. Your assignment of digits to cells must satisfy two conditions: (1) Row 1 plus row 2 must equal row 3 (considering each row as a three-digit number). (2) The digit i must be located immediately next to (above, below, to the right, or to the left) the digit i-1, for $i = 2, . . . , 9$. This second condition means you may place the digit 1 anywhere, but 2 must be placed *next* to 1 along a row or column (not diagonally), 3 must be placed next to 2, and so on. This is what is meant by calling this problem an integer-path problem.

Stop reading and try to solve the problem.

You could begin by trying various hypotheses as to sequences of filling the digits 1, 2, . . . 9 in the nine cells, but this process is long, slow, and chancey. A few inferences that can be made from the above information allow one to greatly reduce the number of hypotheses that must be tested. Probably the most important insight to gain at an early stage in working on the problem is to notice that, of two cells adjacent along a row or column, one must be filled with an odd number and the other with an even number. This arrangement can easily be proved by taking any two cells adjacent along a row or column and considering all possible paths from one cell to the other, realizing that an odd number of steps is required to reach the adjacent cell no matter what path is taken within the matrix. However, it is really not essential to *prove* the theorem that cells adjacent along a row or column must have one

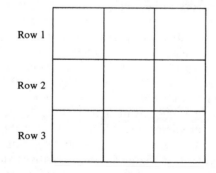

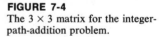

FIGURE 7-4
The 3 × 3 matrix for the integer-path-addition problem.

even number and one odd number. A large number of trial-and-error substitutions executed within a few minutes will show that the theorem is almost certainly true. Stop reading and try to solve the problem, if you could not do so before.

Given this theorem, you can now easily determine that the corner cells and the center cell must be filled with odd numbers and the other four cells with even numbers, since there are five corner plus center cells and five odd numbers. This determination cuts the number of alternatives per cell of the matrix approximately in half, greatly reducing the search space. The restrictions might well be represented in a figure such as Fig. 7-5.

If you have not yet solved the problem, stop reading and try again.

The astute problem solver might infer that in the right-hand column an odd plus an even number is equal to an odd number, but in the middle column an even number plus an odd number is equal to an even

	No carry	Carry	
Row 1	Odd	Even	Odd
Row 2	Even	Odd	Even
Row 3	Odd	Even	Odd

FIGURE 7-5
Restrictions on the digits that can fill cells in the integer-path-addition problem. The restrictions come from considering that the four corner cells and the center cell must have odd digits in them.

number. This event can only happen if there was a carry of 1 from the right-hand column to the middle column. Thus, we know that the sum of the two upper digits in the right-hand column must be greater than 10. Furthermore, in the left-hand column, an odd number plus an even number is equal to an odd number. Thus, there can be no carryover from the middle column to the left-hand column. Hence, the sum of the top two digits in the middle column is 9 or less. Also, we know that the sum of the digits in the left-hand column is 9 or less, since by given information there can be no carry from the left-hand column. With all these restrictions, it is a relatively simple matter to consider the small number of hypotheses that are consistent with these restrictions. Perhaps the easiest way to proceed is to focus on the middle column and test all the hypotheses that are consistent with the information.

This procedure means actually testing only six possible assignments of digits to the top two cells in the middle column—namely, 6-1, 4-3, 4-1, 2-5, 2-3, and 2-1. It is a relatively simple matter to check each of these assignments to see if any path of integers consistent with the assignment could solve the problem. It turns out that only the 2-3 assignment of digits to the top two cells in the middle column will solve the problem. Furthermore, this assignment will solve the problem in only one way, namely, the way shown in Fig. 7-6.

Row 1	1	2	9
Row 2	4	3	8
Row 3	5	6	7

FIGURE 7-6
The solution to the integer-path-addition problem.

Another example of classificatory contradiction combined with inferences based on numerical properties is provided by the *lonesome-eight problem*, which was originated by Chessin (1954):

Determine all of the digits represented by X in the following long division and also determine the remaining four digits of the five-digit answer of which 8 is the third digit, as shown in Fig. 7-7.

Stop reading and try to solve the problem, using the method of contradiction to draw inferences.

```
                8
    X X X ⌐ X X X X X X X X
           X X X
           X X X X
             X X X
           X X X X
           X X X X
```

FIGURE 7-7
The lonesome-eight problem.

Since 8 times the divisor is a three-digit number, we know that the divisor must be 124 or less because $8 \times (125 + z) = 1,000 + 8z$, which is a contradiction for all $z \geq 0$. We can also determine that the last digit of the quotient must be 9, since 8 times the divisor would equal a three-digit number (a contradiction). The initial digit of the quotient must be greater than 7, because 7 times any number less than or equal to 124 would leave a remainder that was greater than a two-digit number when subtracted from the dividend (a contradiction). Now stop reading and try to solve for the rest of the unknown digits, if you did not before.

Since the first digit of the quotient multiplied by the divisor equals a three-digit number, we know it cannot be 9. Thus, the first digit of the quotient must be 8. The second and fourth digits of the quotient must be zero, because in both cases two digits from the dividend were brought down in the work below. So we have the quotient, 80809. Making use of the two places in the work below the dividend in the long division where differences are 99 or less (two-digit numbers), we can determine that 8 times the divisor must be a number between 990 and 999. The only divisor that will multiply by 8 to give a number between 990 and 999 is the divisor, 124. Numbers 123 and less are rejected by contradiction. Thus, we have the quotient and the divisor, and from them we can determine the dividend and all of the values of X in the work underneath the dividend (see Fig. 7-8). Note that in solving this problem, we contradicted large classes of hypotheses (solutions) in making each inference.

```
                    8 0 8 0 9
      1 2 4  | 1 0 0 2 0 3 1 6
                9 9 2
                1 0 0 3
                  9 9 2
                  1 1 1 6
                  1 1 1 6
```

FIGURE 7-8
The solution to the lonesome-eight problem.

ITERATIVE CONTRADICTION IN INFINITE SEARCH SPACES

Occasionally, the method of contradiction can be used in problems that have (initially, at least) an infinite number of possible solutions. Naturally, in such cases, it is necessary to rule out large or infinite classes of alternatives by the method of contradiction. A particularly simple example of the use of classificatory contradiction is afforded by the following problem adapted from Polya (1957).

In numbering the pages of a book, a printer used 3,289 digits. How many pages were in the book, assuming that the first page in the book was numbered 1?

Stop reading and try to solve the problem, using the method of classificatory contradiction to rule out all but one of the infinity of positive integer answers.

Of course, there are not an infinite number of solutions to the problem, once one draws a relatively trivial inference. The inference is that the number of pages cannot possibly be greater than the number of digits (3,289), since at least one digit has to be used to number each page. Thus, we might well regard this problem as an example of classificatory contradiction in a large, but finite, search space. However, since an inference using the method of contradiction was necessary in order to make the search space finite, it seems appropriate to consider the problem as having an initially infinite search space.

The problem provides a particularly simple example of the use of the *iterative* method of taking a preliminary estimate of the goal, determining the magnitude of the error of the estimate from the goal, then moving in the direction of the goal to obtain another estimate along with a magnitude of error, and so on, hoping ultimately to converge upon the goal. Since the preliminary estimates are contradicted by the given information, I think it is useful to consider iterative methods as a subclass of the method of contradiction, a subclass that is particularly useful in solving problems with infinite (but ordered) search spaces. Now stop reading, and try again to solve the problem, if you could not before.

To make the iterative solution to this problem clear, imagine that we start with a preliminary estimate of nine pages for the book. Each page is a single digit. Thus, nine digits in all would be used to number the book. This number is clearly too little, so we move up to the end of the two-digit numbers, namely, the number 99. Numbering 99 pages requires 9 single digits plus 90 two-digit numbers, for a total of 9 + 180, or 189 digits. The number 189 is still substantially below the number 3,289. Thus, for our next estimate we will take the end of the three-digit numbers, namely, 999. Numbering 999 pages uses nine single-digit numbers plus 90 two-digit numbers plus 900 three-digit numbers, for a total of 9 + 180 + 2700, or 2,889 digits. This figure is quite close to the target of 3,289 digits, so we are encouraged perhaps to try a more direct analytic method at this point, namely, subtracting 2,889 from 3,289 to obtain an additional 400 digits that are needed to achieve the goal. (However, note that one could continue to use the iterative method.) Since we are now in the four-digit numbers, each page will require four digits. Thus, to use 400 more digits will require 100 additional pages. Thus, we can infer that we must add 100 to 999 to obtain an answer of 1,099 pages in the book, in order to use up 3,289 digits.

Iterative methods are frequently used in the numerical solution for roots of equations. For example, consider the following problem:

Determine the roots (permissible values of x) that satisfy the equation $x^6 - 4x^5 + 2x^4 + 3x^3 - 7x^2 + 13x - 30 = 0$.

Stop reading and attempt to specify an iterative method of contradiction, by which one might determine each root (permissible value of x) for the preceding equation. You can assume that you have a computer at your disposal to carry out the large number of steps that might be required in order to converge upon each real solution for this equation.

For large enough positive or negative values of x, the x^6 term in the expression must dominate the rest of the expression (be greater than the sum of all the other terms in the expression). Thus, for sufficiently large positive or negative values of x, the expression $x^6 - 4x^5 + 2x^4 + 3x^3 - 7x^2 + 13x - 30$ must be greater than zero and monotonically increasing for more extreme positive or negative values of x. Thus, there can be no real solutions for values of x more extreme than these points at which the expression $x^6 - 4x^5 + 2x^4 + 3x^3 - 7x^2 + 13x - 30$ begins from a positive value monotonically to increase without limit. You can either determine these points or else you can make a safe guess, based on the values of the coefficients of the terms in the expression. In this case, you might assume the function to be monotonically increasing above zero for values of x greater than 1,000 or less than $-1,000$. Assuming that the function is monotonically increasing from positive values beyond this range ($-1,000 \le x \le +1,000$), we know that there can be no zero crossings (roots) beyond the interval from $-1,000 \le x \le +1,000$. Thus, by contradiction, we have ruled out all values of x greater than 1,000 or less than $-1,000$. In the present case, we have not done this in a careful way, but we could. Now stop reading and try to solve the problem again, if you failed to sketch an iterative solution method before.

To determine solutions for x within the interval from $x = -1,000$ to $x = +1,000$, you may define a step size such that you think it unlikely that there would be two different solutions within a single step. In the present instance, we might choose a step size of unity, though if we wish to be more careful we might pick steps of less than unity. Now we can evaluate the function $x^6 - 4x^5 + 2x^4 + 3x^3 - 7x^2 + 13x - 30$ at all integer values of x from $-1,000$ to $+1,000$ to see if the value of the function changes sides (goes from minus to plus or plus to minus) over the step. If the function does change signs, we know that there is a solution (zero crossing) within the interval defined by that step. We can then proceed to use essentially the same method (but more efficiently dividing the remaining interval in half each time) to determine the value of the solution to as fine a degree of approximation as we wish. If the function does not change sign over a step, we assume that there is no solution within that interval.

This numerical method for solving higher order equations is called the *half-interval search technique* and is an excellent example of the iterative use of the method of contradiction. The half-interval search technique uses contradiction, because we consider all intervals over which the function does not change sign not to contain a solution to the equation, since a solution involves a zero crossing (passage from a plus value of the function to a minus value of the function). The absence of a change from plus to minus over some interval contradicts the possibility of a solution in that interval, provided we are justified in assuming that the function cannot have two zero crossings over that interval. If there were some reason to doubt the validity of this for the chosen step size, you could always choose a smaller step size to see if more real roots would be discovered, using the smaller step size. Naturally, if you already found n real roots for an equation of degree n, then you know you have obtained all the roots that it is possible to obtain, since there are at most n real roots for an equation of degree n.

Obviously, the same iterative method can be used to solve other types of equations involving logs, exponents, trig functions, and the like. In addition, you can define multidimensional analogs of the above iterative method to solve sets of several equations with several unknowns. However, it clearly gets more and more time consuming and more and more complicated, the greater the number of unknowns or the more complex the functions.

8

Working Backward

THEORY

The method of *working backward* is similar to the method of contradiction (Chapter 7) and the method of drawing inferences about the goal (Chapter 3) in that all three focus on the goal to a great extent and consider it rather than the givens as the starting point for the problem-solving process. However, working backward differs in the way the goal is considered in relation to the givens. With the methods of contradiction and drawing inferences from the goal, the goal is considered to be part of the given information, and we attempt to derive consequences from the goal in conjunction with the givens. Thus, the direction of inference is from the goal statement to some new statements. In working backward, the goal is not considered to be a piece of given information. We start with the goal, but instead of drawing inferences from it, we try to guess a preceding statement or statements that, taken together, would imply the goal statement. Hence, the direction of inference is the same as in working forward—namely, from the given information to the goal. We start at the end point and try to determine preceding statements, which need not necessarily be given statements but which, when taken together, will produce the goal. Then we try to determine other statements that will imply those statements, gradually working our way back. We hope to arrive at given information that is sufficient to derive everything in between the givens and the goal.

Why should we want to reverse direction like this, proceeding from the goal to the givens rather than from the givens to the goal? When is this method more appropriate than working forward, and why? That is, which problems are appropriate for working backward and which for working forward?

The method of working backward is likely to be useful if a problem satisfies two criteria. The first is that the problem should have a *uniquely specified goal*, as is the case for all proof problems. Whenever there is a single, clearly, and completely specified goal stated in the problem, you should seriously consider the possibility of working backward. This approach is particularly true if, in contrast to the single goal statement, there are large numbers of given statements. Newell, Shaw, and Simon (1962) have clearly stated the advantage to working backward in such problems — namely, there is no ambiguity as to what statement to start with when you work backward, whereas such ambiguity is considerable when you work forward. As they so aptly put it, working forward in such a problem is analogous to finding a needle in a haystack, whereas working backward is analogous to the needle finding its way out of the haystack. You can start from many places outside the haystack in trying to find the single location of the needle. By contrast, the needle starting in a single location can solve the problem of getting out of the haystack by getting to an extremely large number of alternative locations outside the haystack.

In the needle-in-the-haystack problem, the large number of givens have a *disjunctive* relationship to one another. That is, to solve the problem, you need to get from any one of these givens to the goal or by the method of working backward from the goal to any one of the large number of different givens. In many problems to which the method of working backward is appropriate, such as proof problems, the givens have a *conjunctive* relationship to one another. That is, you must use several of the givens to derive the goal. Thus, in the method of working backward, it will usually be necessary to work backward from the goal to get to several of the givens rather than to only one of the givens. Nevertheless, the method is frequently very useful in such problems, because the unique starting point so frequently directs you to just those aspects of the given information that are relevant to the solution.

In problems where the goal is not so clearly and completely specified and there are, in fact, a variety of possible alternative goals, the advantages of working backward are often largely eliminated. In the case of inference problems (problems with nondestructive operations), the method of working backward is suitable for proof problems but not generally for find problems. In the case of action problems (problems with destructive operations), essentially the same distinction applies — namely, problems with uniquely specified goals are appropriate for the method of working backward and problems in which only certain characteristics of the goal are specified are generally not so suitable.

The second criterion of a problem as to whether the method of working backward is applicable concerns the nature of the operations specified in the problem. If all operations are unary and one-to-one operations, the method of working backward is likely to be helpful. *Unary operations* are operations that take one given input statement and produce one output statement. *One-to-one operations* are operations for which it is possible to uniquely determine what input statements produce the output statement. (These concepts will be discussed in more detail in Chapter 10.) Since a well-defined unary operation yields a unique output statement for each input statement, there is no ambiguity concerning the result of an operation when applied to some state working forward. However, a well-defined unary operation need not be one-to-one; several different input statements could produce the same output statement. Thus, when operations are not one-to-one, working backward may lead to a more rapidly branching tree of possible action sequences than will working forward. In such cases, it would generally be preferable to work forward.

The unary property of an operation is not as important as the one-to-one property for the applicability of working backward. Binary or ternary operations take two or three input statements and produce one output statement. Using the method of working backward, it would be necessary, given the output statement, to produce two or three input statements. Superficially, this might seem similar to what happens using the method of working backward when operations are not one-to-one. However, there is an important difference — with binary or ternary operations, the inference process is essentially as complicated when working forward as when working backward. A conjunction of inputs is related to a single output statement with binary or ternary operations, and this is equally true working forward or backward. From one problem to another, it may be more or less difficult to work in the forward or the backward direction, but the existence of binary or ternary operations does not necessarily invalidate working backward, and examples of working backward in just such problems will be given later in the chapter. In such problems, working backward generally results in a set of subgoal statements, and then the subgoal statements are frequently derived from the givens by working either forward or backward.

By contrast, with operations that are not one-to-one, working backward generates a multiplicity of alternative input statements that are disjunctively related one to another (rather than conjunctively related). Thus, you are generating a large set of alternative prior statements, using the method of working backward, which would never be present using the method of working forward, since only one of these statements is necessary in order to derive the goal (not a conjunction of several or all of them). Thus, the more critical property of an operation that is beneficial for working backward is the one-to-one property, while the unary property facilitates problem solving working either forward or backward.

Another way to state the critical one-to-one property of operations desirable for working backward is to say that the operations in the problem should admit of the possibility of defining inverse operations. *Inverse operations* are operations that go from the output statement back to the input statement and reverse the effect of some given operation. Whenever operations are one-to-one, it is possible to specify well-defined inverse operations that will uniquely produce the input statement from the output statement. Clearly, if the originally specified operations were not one-to-one, a multiplicity of different input statements would produce the same output statement, and you would have no way of defining an inverse operation that uniquely specified a single input statement that produced some output statement.

If either action or inference problems specify a single goal and if the operations specified in the problem are one-to-one (admit the definition of inverse operations), working backward will often, though not always, be preferable to working forward. However, where one or both of these criteria are not satisfied by a problem, working backward will likely be inferior to working forward. Hence, working backward is by no means universally preferable to working forward. In fact, i' my experience, it is generally more difficult to work backward than to work forward. Nevertheless, there is a large class of problems to which the method of working backward is appropriate, and some examples are given in the following section.

APPLICATIONS

Action Problems

The following clever little *doubling-game problem* illustrates the usefulness of working backward:

> Three people play a game in which one person loses and two people win each game. The one who loses must double the amount of money that each of the other two players has at that time. The three players agree to play three games. At the end of the three games, each player has lost one game and each person has $8. What was the original stake of each player?

Offhand, it seems as if there is insufficient information to determine the answer. However, because the players all finish with the same amount of money, $8, it is possible to compute their original stake by working backward. We will label the first losing player P_1, the second P_2, and the third P_3. Stop reading and try to solve the problem, if you did not do so before.

At the end of game 3, P_1, P_2, and P_3 each had $8. Working backward to the end of game 2, P_1 must have had $4 and P_2 $4, since both won in game 3 (P_3 lost), and thus both had their stakes doubled by the results of game 3. Since P_1 and P_2 each gained $4 in game 3, P_3 must have lost $8 in game 3, so P_3 had $16 at the end of game 2. Now work backward to determine the stakes of each player at the beginning of game 1, if you did not solve the problem before.

The complete solution obtained by working backward is shown in Fig. 8-1, where we observe that in the beginning P_1 had $13, P_2 had $7, and P_3 had $4.

Note that if the players did not all end with the same amount of money, it would be impossible to determine what each player started with, because the order in which the players won and lost would make

	P_1	P_2	P_3
End of game 3	$ 8	$ 8	$ 8
End of game 2	4	4	16
End of game 1	2	14	8
Beginning	13	7	4

FIGURE 8-1
Working backward to solve a doubling-game problem.

a difference. However, in the present instance, the order in which the players won or lost games makes no difference to determining the initial stake of each player.

Also, if you had names for the players, you could not tell which player started with $13, which with $7, and which with $4, unless you know the order in which they won. Here I simply named the first losing player P_1, the second P_2, and the third P_3. This was completely adequate, since the stated goal did not require pairing stakes with named players.

This doubling-game problem is an extreme example of the usefulness of working backward, since it is essentially impossible to solve the problem, except by working backward. The reason is that there is a uniquely defined goal, but no given state is specified at all. In fact, the problem is to derive a given state that, in conjunction with the operations, will produce the goal state. Although the operations are stated in a forward direction, they easily admit the definition of unique inverse operations (in which two people have their stakes cut in half and the other person has his stake increased by the sum of the amounts the others were decreased). Thus, it is clear we must use the method of working backward to solve this problem. Of course, it is equally correct to say that what we have done is to transform the operations into inverse operations and reverse the goal and the givens, taking the goal as the givens and attempting to derive the given state from the goal. Obviously, it makes little difference which way we describe what was done in this problem, since what was done is precisely the same under either description.

Nim games are games in which each player takes away tokens subject to a variety of restrictions and tries to be the last—or not the last—to take a token. Nim games provide excellent examples of the usefulness of working backward to determine optimal strategy. One example is the following:

> Fifteen pennies are placed on a table in front of two players. Each player is allowed to remove at least one penny but not more than five pennies at his turn. The players alternate turns, each removing from one to five pennies n number of turns, until one player takes the last penny on the table, and wins all 15 pennies. Is there a method of play that will guarantee victory? If so, what is it?

Stop reading and try to determine the optimal strategy by working backward.

If you conjecture yourself to be in the goal state, this state would clearly consist of being the player whose turn it is to move, and there being anywhere from one to five pennies on the table. In this state, you could take all of the pennies left on the table and be the winner. Now, can you work backward from this set of possible goal states to conjecture a preceding state for the other player in which, no matter what that player does, you will be in one of these desirable goal states? Stop reading and try to solve the problem, if you did not before.

It is clear that if you confronted the opposing player with six pennies on the preceding turn, no matter how many pennies he or she took (from one to five, there would still be from one to five pennies on the table when it was your turn, giving you the victory. Thus, you should try to confront your opponent with six pennies after your move. But you cannot do this on your first move, so you must work backward again and ask what previous position you would have to put your opponent in so that, no matter what he did, you could have six pennies on the table after your move. Now stop reading and try to solve the rest of the problem, if you did not before.

Some thought reveals that if you could confront the opposing player with 12 pennies after your preceding move, then no matter how many pennies he took (from 1 to 5), you would be able to take enough pennies to confront him with 6 pennies on the next turn. Thus, you want to confront your opponent with 12 pennies, and you can do that on your first move by removing 3 pennies from the board.

Note that in this nim problem, there was no uniquely defined goal state, since the goal is to take the last one, two, three, four, or five pennies on the table, and we cannot know which of these moves would constitute the goal in any nim game we might win. Of course, you could easily transform this into the unique goal of facing your opponent with a table having zero pennies on it on the other player's next move. Obviously, it makes no difference which way you look at this nim problem. The point is that working backward will frequently generate many possible preceding states, and this fact does not necessarily invalidate the method of working backward. In the present instance, working backward two steps yields a unique preceding number of pennies that you should confront your opponent with on the turn prior to your opponent's last turn — namely, six pennies. Thus, considering only your own sequence of moves (rather than all the different moves that might be made by your opponent), we see that the method of working backward in the present problem proceeds to get one preceding state from one succeeding state. Thus, working in the backward direction, the tree of possible states is continually being pinched back to a single state at every alternate move. The lesson in this problem is that you should not be too easily discouraged from working backward by a multiplicity of preceding states, if this multiplicity is only a temporary phenomenon or a one-time sprouting of branches of the tree followed by no further increase. We cannot expect the method of working backward to always produce a single one-to-one chain of states back to the givens from the goal with no alternatives to investigate.

Working backward was preferable to working forward in the preceding problem because the number of different action sequences that had to be considered working backward was considerably smaller than the number that had to be considered working forward. You could, at every alternate move, determine a unique state in which you should be. This is the most general criterion for the applicability of working backward, namely, that it produce a smaller space of alternative action sequences than would be produced by working forward. Sometimes working backward is preferable to working forward because it produces a smaller set of action sequences to consider when combined with a hill-climbing approach. An example is provided by the following *checker-rearrangement problem*:

> On an infinitely extended checkerboard, one is given three black checkers and two white checkers initially placed in immediately adjacent squares on a single row, proceeding from left to right, as shown in Fig. 8-2: black *(B)*, white *(W)*, black, white, black. The problem is to transform this arrangement of alternating black and white checkers into an arrangement in which all three black checkers are on the left and both white checkers are on the right *(BBBWW)*, with all checkers being in adjacent squares and in the same row (see Fig. 8-2). The allowable operation is to move two adjacent checkers at a time, one of which must be a black checker and one a white checker. During a move, the two checkers being moved must remain together at all times, with no reversal of their left-to-right order. You are permitted to move a white-black or black-white pair of checkers to any adjacent pair of unoccupied squares along the same line. Note that there is no need to keep the checkers that are not being moved in immediately adjacent squares at any time. That is, there may be unoccupied squares between checkers at various stages between the givens and the goal. Also note that the five checkers in the goal state need not occupy the same five squares on the checkerboard as they did in the given state. They may occupy any immediately adjacent five squares in the same row.

Stop reading and try to solve the problem by defining an evaluation function and then using the method of working backward in conjunction with a hill-climbing approach on this evaluation function.

Given   Goal

FIGURE 8-2
Given and goal states for the checker-rearrangement problem.

This problem is frustrating to solve by working forward because there are many possible moves at each point, and it is not at all clear how to hill-climb in order to get to the goal. Nor is it clear what subgoals you ought to set on the way to the goal statement. By contrast, working in the backward direction, there is only one pair of checkers that can be moved initially, namely, the third and fourth checkers in the row. After the first move, the number of possibilities at each move is also more limited than would happen if you moved in the forward direction. The solution is relatively easy to obtain working backward because of this much greater restriction in the number of possible initial moves. Now stop reading and try to solve the problem, if you did not do so already.

In addition to working backward, it is helpful to define as an evaluation function the number of immediately adjacent black-white and white-black pairs of checkers. This evaluation function has a value of 1 in the goal state and a value of 4 black-white or white-black transitions in the given state. Choosing actions that increase this evaluation function is of some assistance in narrowing the space of possible moves in working backward from the goal state to the given state. Of course, it is also of some help in working forward. However, the problem is considerably easier working backward, because of the greater restriction of initial moves from the goal state, as well as for psychological reasons peculiar to this problem. An optimal solution to the problem, along with the values of the evaluation function for each state, is shown in Fig. 8-3.

Because of the greater restriction of initial moves starting from the goal and working backward than starting from the givens and working forward, working backward was clearly indicated in this checker-rearrangement problem. However, even if there is no reason to prefer the method of working backward in a problem, you should always consider its use whenever there is no reason to prefer working forward. That is, there are many problems in which it may not be obvious *a priori* which method, working backward or working forward, is superior; in such cases, you might well try working forward and, if it did not seem to be working out well, then try working backward.

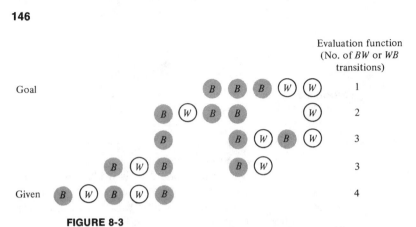

FIGURE 8-3
Working backward to solve the checker-rearrangement problem.

As a final example of the method of working backward, let us consider a *water jar problem*:

> Given a jar that will hold exactly 7 quarts of water, a jar that will hold exactly 3 quarts of water, no other containers holding water, but an infinite supply of water, describe a sequence of fillings and emptyings of water jars that will result in achieving 5 quarts of water.

Stop reading and try to solve the problem, working backward.

Obviously, at the goal state we will have 5 quarts of water in the 7-quart jar. There are several ways this might be achieved from a preceding state, working backward. First, we might have 2 quarts of water in the 7-quart jar and pour in 3 quarts from the 3-quart jar. Second, we might have 3 quarts of water in the 7-quart jar and pour in 2 quarts from the 3-quart jar (this seems less likely than the first alternative). Third, we might have 4 quarts of water in the 7-quart jar and pour in 1 quart from the 3-quart jar. Fourth, we might have 7 quarts in the 7-quart container, 1 quart in the 3-quart container, and pour off 2 quarts into the 3-quart jar. Fifth, we might have 6 quarts in the 7-quart jar, 2 quarts in the 3-quart jar, and pour off 1 quart into the 3-quart jar. Now stop reading, and try to solve the problem, if you have not done so already.

Of the five alternatives for the state preceding the goal, the first and the fourth are the most plausible, since they involve quantities of water in one or the other jars that are easy to achieve—namely, 7 quarts in the 7-quart jar and 3 quarts in the 3-quart jar. Thus, we might well confine our attention to these two possibilities, at least initially. Now stop reading and try to solve the problem, if you have not done so already.

Although it is possible to achieve 2 quarts in a 7-quart jar as a subgoal and fill up the 3-quart jar as specified in the first alternative, the fourth alternative is actually optimal. Working backward from the fourth alternative (7 quarts in a 7-quart jar and 1 quart in the 3-quart jar), we set as the subgoal the achievement of 1 quart in either jar. It is probably not particularly useful to continue working backward any longer after having defined these five alternative preceding states. Rather, in attempting to achieve any one of these states, such as 7 quarts in the 7-quart jar and 1 quart in the 3-quart jar, it is probably most useful to set this as a subgoal and work forward from the given information in order to achieve it. In the present instance, it is quite easy to achieve the state of having 7 quarts in the 7-quart jar and 1 quart in the 3-quart jar. To achieve 1 quart in the 3-quart jar, fill the 7-quart jar, pour off 3 quarts two successive times to achieve 1 quart in the 7-quart jar, then transfer this 1 quart to the 3-quart jar. Now fill up the 7-quart jar, and the subgoal is achieved. After this, of course, it is simple to pour off 2 quarts from the 7-quart jar into the 3-quart jar, which already contains 1 quart. This leaves 5 quarts in the 7-quart jar, which is the goal.

This problem nicely illustrates how working backward can permit you to define a subgoal, which you can then achieve by working forward. This pattern is typical of working backward in both action problems such as this one and the inference problems to be discussed next.

Inference Problems

As a simple example of the method of working backward in *inference problems*, consider the following proof problem:

> If $A > 0$ and $B > 0$, then $A^2 - AB + B^2 > 0$. More general versions of this theorem are true, but it needlessly complicates the example of the method of working backward to consider these more general versions. Thus, we shall restrict our consideration to the case where $A > 0$ and $B > 0$.

Stop reading and try to solve the problem by working backward.

To apply the method of working backward, we first state the conclusion $A^2 - AB + B^2 > 0$. A preceding statement that would imply that conclusion can be obtained by factoring the expression $A^2 - AB + B^2$ into $A(A - B) + B^2 > 0$. If we could show this expression to be true, then it would imply the desired conclusion. Stop reading and try to solve the problem, if you did not do so before.

By working backward we note that we could derive the expression $A(A - B) + B^2 > 0$ from three previous expressions: first, $A > 0$, which is given information; second, $B^2 > 0$, which is true for all real numbers including B; and, third, $A - B > 0$. We cannot derive $A - B > 0$ from the given information, but we could just assume it as one case. Obviously, in some cases where both $A > 0$ and $B > 0$, A will be greater than B. So in the case where $A > B$, then $A - B > 0$, and the theorem is proved. Stop reading and try to solve the problem, if you did not before.

Now, work backward from the goal statement $A^2 - AB + B^2 > 0$ to try to derive the goal expression in the case where $B > A$. In this case, we factor the goal expression into $A^2 + B(B - A) > 0$, which will be true if, first, $B > 0$ (given information); second, $A^2 > 0$ (true for all real A); and, third, $B - A > 0$. Now, $B - A > 0$ follows from $B > A$. Thus, we have established that the conclusion follows where $A > B$ or $B > A$. In addition, we have to show that the conclusion follows when $A = B$, but this matter is trivial. This problem is so short and relatively trivial that many people may not notice how much they are working backward in solving it. However, a careful examination reveals that the critical insights come from focusing on the goal statement and noticing what it can be factored into.

Frequently we use the method of working backward for only a few steps in order to derive some more congenial formulation of the statement to be proved, then we proceed to work in a forward direction in trying to derive this more congenial formulation. Working backward may simply result in the substitution of a single subgoal (directly related to the goal) for the original goal, or it may result in the substitution of two or more subgoals in place of the original goal.

An extremely simple proof of the Pythagorean Theorem can be obtained by initially working backward from the algebraic goal statement to obtain a single, more geometric subgoal.

> As you may remember, the Pythagorean Theorem states that, for any right triangle, $c^2 = a^2 + b^2$, where c is the length of the hypotenuse. Prove this theorem, where the givens are, first, the axioms of Euclidean geometry; second, the definition of the area of a rectangular figure (length times width); and, third, the assumption that the areas of nonoverlapping figures are additive.

Stop reading and try to prove the theorem by working backward to obtain a single subgoal.

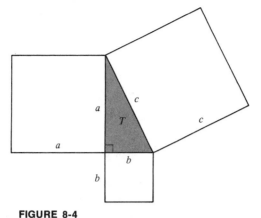

FIGURE 8-4
Squares erected on the sides of a right triangle
by working backward one step from the algebraic
formulation of the Pythagorean Theorem, $c^2 = a^2 + b^2$.

Instead of trying to use the givens in an attempt to derive the goal expression, it is far simpler to look at the goal expression and note that it is asserting that the area of a square with side c erected on the hypotenuse is equal to the sum of the areas of the squares with sides a and b, respectively, erected on the other sides of the right triangle. This situation is shown graphically in Fig. 8-4.

Thus, by working backward from the goal expression $c^2 = a^2 + b^2$, we have obtained a subgoal that might prove more tractable than the original goal—namely, to show that the area of the large square with side c is equal to the sum of the areas of the smaller squares with sides a and b. Now stop reading and try to prove the subgoal, if you have not done so already.

To prove that the subgoal statement is true, you need to get expressions for the areas of the three squares that are in the same terms, so that you can determine whether the sum of the two smaller areas equals the largest area. Since the original triangle is the basis for any relation among the areas of these three squares, it seems natural to try to express the area of each square in terms of the area of the original triangle, T. Now stop reading and try to formulate this expression, if you have not done so already.

It is quite straightforward to reflect the triangle, T, onto the squares erected on the nonhypotenuse sides, a and b. Assuming that $a > b$, we can lay out two triangles on the square with side a and have a rectangle with area $a(a - b)$ left over in the square. In the case of the smaller square with side b, two T triangles will use up a rectangle that has an area greater than the area of the square by an amount equal to a rectangle with area $b(a - b)$. All this is shown in Fig. 8-5. Thus, we can replace the terms $a^2 + b^2$ by $4T + a(a - b) - b(a - b) = 4T + (a - b)^2$.

Figuring out how to lay out T triangles inside the largest square (with side c) is more of a challenge. However, with the idea of reflecting the original triangle about the side it shares with the various squares erected on its side, we should eventually wind up laying out four T triangles inside the square with side c, as shown in Fig. 8-5. To lay out the four T triangles within the largest square, the most critical property of the original triangle to note is that $\alpha + \beta = 90°$. Taking four T triangles out of the square with side c leaves a square with side $(a - b)$ inside the four triangles. Then the area of the large square is $c^2 = 4T + (a - b)^2$, exactly what was obtained for the sum of the areas of the other two squares. Thus, the area of the square erected on side c is equal to the sum of the areas of the squares erected on sides a and b, and the Pythagorean Theorem is proved.

An example of working backward to find several subgoals is provided by the following proof problem:

> You are given the following four assumptions: (1) Multiplication is commutative; that is, $AB = BA$. (2) Equals added to equals are equal; that is, if $A = A'$ and $B = B'$, then $A + B = A' + B'$. (3) The left distributive law applies; that is, $C(A + B) = CA + CB$. (4) The transitive law applies; that is, if $A = B$ and $B = C$, then $A = C$. From these four givens, prove the right distributive law – that is, $(A + B)C = AC + BC$.

Stop reading and try to solve the problem, working backward to derive subgoals.

To prove this theorem, a good beginning point would be to start at the goal statement $(A + B)C = AC + BC$, and write down one or more preceding statements from which you could derive the goal statement as a conclusion. One pair of preceding statements from which you could derive the goal statement is the following: $(A + B)C = X$ and $X = AC + BC$. From these two preceding statements you could derive the goal statement, using the transitive law. Thus, we have subdivided the goal into two subgoals that are, however, somewhat dependent one upon another in that we must try to transform each of the expressions that are considered to be equal in the theorem into expressions that are identical (indicated by X). Stop reading and try to solve the problem, if you did not before.

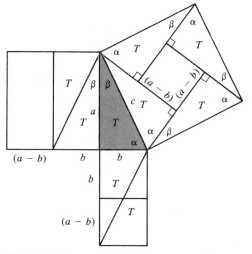

FIGURE 8-5
Expressing the areas of all three squares in terms
of T (the area of the original triangle), a, and b,
so that c terms are eliminated from the expression
for the area of the large square.

In this problem, just applying legitimate operations to the expressions $(A + B)C$ and $AC + BC$ will easily result in deriving expressions that are identical in each case. So $(A + B)C = C(A + B)$, by the commutative law for multiplication. $C(A + B) = CA + CB$, by the left distributive law. Therefore, $(A + B)C = CA + CB$, by the transitive law of equality. Now $CA = AC$ and $CB = BC$, by the commutative law for multiplication. Therefore, $CA + CB = AC + BC$, because equals added to equals are equal. Thus, $(A + B)C = AC + BC$, by the transitive law for equality, and the theorem is proved.

9

Relations Between Problems

When mankind has a satisfactory theory of problems, it will be possible to state many deep and detailed relations between different types of problems. But even without such a theory, we can still state certain basic types of relations between different problems.

In particular, five fundamental types of relations can obtain between two problems, *a* and *b*: First, problem *a* is unrelated to problem *b* (problem *a* and problem *b* have no common elements). Second, problem *a* is equivalent to problem *b* (*a* and *b* have the same problem elements, *a* and *b* are completely analogous, *a* and *b* are isomorphic). Third, problem *a* is similar to problem *b* (problems *a* and *b* have some common elements, problems *a* and *b* are partially analogous). Fourth, problem *a* is a special case of problem *b* (problem *a* is included in problem *b*). Fifth, problem *a* is a generalization of problem *b* (problem *b* is included in problem *a*). When problems *a* and *b* are similar, they may be of approximately equivalent difficulty, *b* may be simpler than *a*, or *b* may be more complex than *a*.

EQUIVALENT PROBLEMS

In determining whether any of these five relations holds between two problems, it is important to note that the critical problem elements concern the types of operations and the relations that can obtain between different expressions or things, not the specific expressions or things themselves. For example, in the checker-rearrangement problem described in Chapter 8, it would make no difference in the elements of that problem, from a problem-solving viewpoint, if the black checkers were changed to quarters and the white checkers were changed to pennies, provided that all the same restrictions and operations still applied. Alternatively, we could replace the black checkers by red poker chips and the white checkers by blue poker chips, and, if everything else remained the same, the problem would be equivalent to the original problem. Similarly, in the nim problem of Chapter 8, in which from one to five pennies were removed by a player on each turn, the pennies could be replaced by any token such as marbles, poker chips, buttons, or stones. In the Tower of Hanoi problem in Chapter 6, the disks of decreasing size could be replaced by any set of tokens having a simple order relation among them.

Problems that differ only with respect to the names attached to different elements of the problem, but all of whose relations and operations are identical, are considered *equivalent*, meaning they are completely analogous or isomorphic. Recognizing that two problems are equivalent may sometimes involve realizing that many of the implied properties indicated by the different names attached to corresponding elements in the two problems are completely irrelevant to the solution of the problem. However, it is usually relatively trivial to recognize such irrelevancies of the differing properties of corresponding elements, and consequently the recognition of equivalent problems is frequently trivial.

SIMILAR PROBLEMS

Two problems can be extremely similar and yet not be equivalent in every respect. For example, in the Tower of Hanoi problem, you may start with 5, 6, 7, or 10 disks on a single spike; the number of disks you begin with is in no way critical to the method of solving the problem as outlined in Chapter 6. You can, in fact, state a solution to the general Tower of Hanoi problem in which n disks must be transferred from one spike to another. Thus, any particular Tower of Hanoi problem (a problem involving the transfer of some particular number of disks from one spike to another) is a special case of the general Tower of Hanoi problem. Any two special cases of the general Tower of Hanoi problem are similar problems, though I would hesitate to call them equivalent problems.

Similarly, in the nim problem, involving one to five pennies that can be removed at each turn, you can construct a large variety of problems that differ in the maximum number of pennies that may be removed on each turn or the number of pennies originally placed on the table. Each of these problems can be solved by essentially the same problem-solving method, though the specific number of pennies that the player should take on each turn will differ from problem to problem. Again, all of these problems are extremely similar, but not completely equivalent.

The preceding two examples of extremely similar problems were examples in which the problems differed only in the quantities of certain elements of the problem. In each case, all of the qualitative or structural characteristics of the problems were identical.

Equivalent Difficulty

There are other partly analogous problems in which the structure is somewhat different in the two problems being compared but still highly similar. For example, consider the following problem called the *fox, goose, corn problem*:

> A man (M), a fox (F), a goose, (G), and some corn (C) are together on one side of the river (straight line) with a boat (B), as illustrated in the given state of Fig. 9-1. The goal is to transfer all of these entities to the other side of the river by means of the boat, which will carry the man and one other entity. The fox and the goose cannot be left alone together, nor can the goose and the corn.

Stop reading and try to solve the problem by recalling the methods used to solve a similar (partly analogous) problem.

Given	Goal	FIGURE 9-1
M F G C B	————————	Given and goal states for the fox, goose,
————————	M F G C B	corn problem.

The fox, goose, corn problem is similar to the missionaries-and-cannibals problem discussed in Chapter 5. Now stop reading and try again to solve the problem, if you did not before.

The solution to this problem is shown in Fig. 9-2. You will note that at one critical stage you must make an apparent detour in order to solve the problem, just as in the missionaries-and-cannibals problem. You might well have suspected that something of this character would be

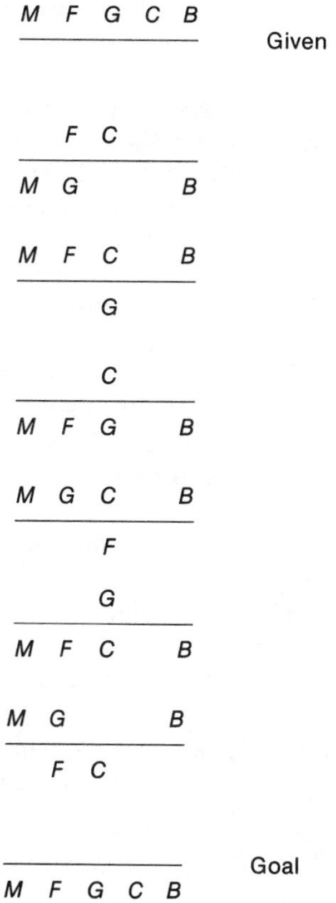

```
M   F   G   C   B
_____     Given


    F   C
_____
M   G           B

M   F   C       B
_____
        G


        C
_____
M   F   G       B

M   G   C       B
_____
        F


        G
_____
M   F   C       B

M   G           B
_____
    F   C


_____     Goal
M   F   G   C   B
```

FIGURE 9-2
Solution to the fox, goose, corn problem.

required, since the two problems are similar in having restrictions regarding what entities can be on the same side of the river with what other entities at the same time. Note, however, that instead of two types of entities, as in the missionaries-and-cannibals problem, this problem has four types. Furthermore, there is only one example of each entity, whereas in the missionaries-and-cannibals problem, any one of the missionaries or cannibals could row the boat. With all these differences, it is surprising that the one primary similarity is nevertheless the dominant element of the problem with respect to problem-solving methods: In each case, we find an implicit evaluation function in terms of the number of entities on the goal side of the river and take actions that increase that evaluation function. In each case, it is necessary to make a brief detour in terms of that evaluation function in order to solve the problem.

A spatial reasoning problem similar to a problem previously considered in this book is the following:

> You are given six coins arranged in two rows (as shown on the left side of Fig. 9-3) so that each coin touches the coins immediately above or below it and to the left or right of it. Specify a procedure for moving exactly two coins so as to achieve the hexagonal arrangement shown on the right side of Fig. 9-3.

Stop reading and try to solve this problem by considering a previously solved related problem.

The most closely related problem is the bowling-pin reversal problem discussed in Chapter 7. Both problems involve spatially distributed objects that must be rearranged in a minimum number of moves to achieve some new configuration. Stop reading and try again to solve the problem, if you did not before.

By analogy to the bowling-pin reversal problem, it is useful to ask which four of the coins will remain in the same position and which two will be moved, going from the given to the goal. Since there are only $(6 \times 5)/2$, or 15 combinations of two "moved" coins, it would be relatively simple to investigate all the possibilities, using the method of contradiction.

However, if you recall that in the bowling-pin reversal problem an effective strategy was to look for maximum subgroups in the given and goal states that occupy the same relative positions to one another, then this perceptual strategy can be applied to quickly give the answer

FIGURE 9-3
Coin-rearrangement problem.

to the present problem. Now stop reading and try to solve the problem, if you have not done so already.

Clearly, the coins in positions 1, 2, 4, and 6 are in precisely the same relative configuration to one another as the top four coins in the goal state. Thus, you can achieve a solution by moving coins 3 and 5 to the two bottom positions. A symmetrically opposite solution can be achieved by keeping coins 1, 5, 6, and 3 in the same positions (forming the bottom of the goal hexagon) and moving coins 2 and 4 to the top two positions of the goal. However, these moves are the only two of the 15 alternative moves of two coins that solve the problem.

Recall the one-heavy-coin problem discussed in Chapters 3 and 5. In that problem, we had to determine which of 24 coins was heaviest, using a beam balance. Obviously, similar principles of problem solving are likely to be involved in problems where the original set of coins is some number other than 24. Furthermore, it is immediately apparent that making the odd coin lighter than the normal coins, rather than heavier, would not change the method of solution in any respect. That is, the one-light-coin problem is equivalent to the one-heavy-coin problem.

Simpler Problems

What happens when we know the odd coin is either heavier or lighter than the normal coins but do not know which of the two relations the odd coin has to a normal coin? This problem is obviously rather similar to the previous problems, but it is different in a much more profound respect than simply a variation in the number of coins of the original set or the heavy versus light nature of the odd coin. In this new problem, where the heavier versus lighter nature of the odd coin is ambiguous, one principle of problem solving still applies—that is, you still get an answer to a three-way question from the balance beam. However, the logical character of the reasoning, given the different outcomes on the balance beam, is much more complicated.

Nevertheless, noticing the analogy to these other problems would enormously aid solution of the heavier-lighter-coin problem. Even if you had not previously solved the one-heavy-coin or one-light-coin problem, it might still be good strategy to pose and solve either of these simpler problems before you attempt to solve the more complex problem.

This strategy of posing similar, simpler problems before working on a complex problem is very useful, since many of the methods of representing information or of solving the problem are common to both. To be sure, the complex problem will invariably have some additional complications. However, if when solving the simpler problem you discovered some of the methods for solving the complex problem, it will be easier to discover the remaining methods of solving the more complex problem than if you had to solve the complex problem as a whole.

Along the same lines, when you originally confronted the one-heavy-coin problem with a set of 24 coins, you might have tried solving a simpler problem, with the odd coin embedded in a set of only three or four coins. With a set of three coins, you have the best opportunity to realize that the balance beam can provide a three-way partitioning of the set of alternatives.

Often, as in the heavy-coin problems, you judge the simplicity of the problem by the number of different elements or complications in it. In the one-heavy-coin problems, which differ only in the total size of the original set of coins, the complication of the problem seems to be variable in a simple way, namely, the change in the number of coins in the total set.

However, we have also noted that a problem in which the odd coin might be either heavier or lighter was a substantially more complicated problem than the problem in which it was known definitely which weight relation the odd coin had to a normal coin. Simplicity in a problem is by no means a simple quantitative concept.

Another problem similar to the one-heavy-coin problem, but simpler to solve because it has one less restriction, is the problem of *three-way-question information theory*. The problem is to determine which element is the unique element, in a set of n possible elements, by successively partitioning the set into three subsets, then ask which of the three subsets contains the unique element.

In ordinary (two-way-question) information theory, the optimal strategy is to divide the total set into two equal or nearly equal parts and to continue dividing the remaining set into two equal parts until the unique element is determined. Interestingly, in three-way-question information theory, the optimal strategy is not always to divide into three equal (or nearly equal) parts, though this is not a bad strategy. If the objective is to minimize the expected number of questions to be asked in order to determine the unique element, then other kinds of partitions besides the equal partition are optimal for some set sizes. For example, if there are six elements in the original total set, the optimal strategy with three-way-question information theory is not to divide into 2, 2, and 2 on the first partition. Instead, the optimal strategy is to divide into 3, 2, and 1, because one-sixth of the time this will give you the answer in one question and five-sixths of the time it will give you the answer in two questions. By contrast, the 2-2-2 split will give you the unique element in two questions in every case. Similarly, if there were seven elements in the original set, you should divide into 3-3-1 rather than 3-2-2—and so on.

Posing and solving the somewhat simpler three-way question information-theory problem provides the surprising information that dividing into equal thirds is not necessarily the optimal solution. This principle can be applied in some cases in the one-heavy-coin problem, though it is limited by the restriction of having two equal subsets in that problem.

Sometimes when you pose a simpler problem you lose all the difficult aspects of the original problem. In that case, solving the simpler problem provides no help at all in solving the original, more complex problem. For example, in the one-heavy-coin problem involving 24 coins, if you chose to investigate the simpler problem involving only 2 coins, you would draw the same type of wrong conclusion regarding the weighing operation that many people fall into when working on the original problem—namely, dividing in half.

More serious than the danger that posing simpler problems will lose the complexity of the original problem is the danger of posing apparently simpler problems that are really more difficult to solve than the original problem. Although it is generally true that reducing the number of elements in a problem reduces the complexity of the problem, it is not always so. Sometimes reducing the number of elements of a particular kind in a problem, or eliminating some of the features of the problem, results in a problem that is more difficult to solve than the original problem. Sometimes the supposedly simpler problem is impossible to solve. The following coin-weighing problem illustrates the danger involved in posing simpler problems:

> You have 10 stacks of quarters with 10 quarters in each stack. One entire stack is composed of quarters, each of which weighs 2 grams less than it should. You know the correct weight of a quarter. You may weigh the coins on a pointer scale, which tells you how many grams a set of objects placed on it weighs. What procedure will determine the light stack in the smallest number of weighings?

Stop reading and try to solve the problem.

This coin-weighing problem is different from the previous coin-weighing problems, primarily because the weighing operation is different. That is, this problem uses a pointer scale, whereas previous problems used a beam balance. A weighing operation using a beam balance provides an answer to a three-way question, but a single weighing on a pointer scale provides an enormously greater amount of information (limited only by the accuracy of the pointer scale). Because of the great difference in the amount of information provided by the pointer versus beam-balance weighings, there is virtually no similarity between the solutions to these two types of coin-weighing problems. Thus, these two types of coin-weighing problems are not really related at all, in the problem-solving sense. The presence or absence of concrete similarities, such as two problems being both concerned with coins and weighing operations, is less important than more abstract similarities concerned with the relationships among givens or between givens and operations. If you were misled into trying to apply similar methods to those used in solving the previous coin-weighing problems, stop reading and try again to solve this problem.

Another approach that fails is to try to simplify the problem by reducing the number of coins in each stack to one coin and simply determining which of the 10 remaining coins is light. This actually makes the number of weighings vastly greater than in the original problem, where one had 10 stacks of 10 coins each. What other way is there to simplify the problem? Stop reading and try again to solve the problem, if you did not before.

The other obvious way to simplify the problem is to reduce the number of stacks. Solving a problem with a reduced number of stacks could facilitate solution of the problem. Now stop reading and try to solve the problem, using this method of simplification, if you did not solve it before.

The simplest problem that can be posed, reducing the number of stacks, is to decide which of two stacks is light. Evidently, this could be done in one weighing by weighing a single coin from one of the two stacks and determining that it is either the correct weight for a quarter or 2 grams less than the correct weight. Of itself, this solution to the two-stack problem does not indicate how you should solve the 10-stack problem. However, it does provide you with a basic familiarity regarding the nature of the information provided by the pointer scale. You should then try to solve a three-stack problem. Now stop reading and try to solve the three-stack problem in a way that will allow generalization to a 10-stack problem, if you have not solved the problem so far.

The three-stack problem can be solved in a single weighing, as can the 10-stack problem. However, an insight is required to accomplish this. I know of no general problem-solving method that would automatically provide you with the critical insight. Attempting to solve the simpler three-stack problem makes it more likely that you would achieve that insight, but it does not guarantee it. If you have not yet solved the problem, stop reading and try to determine what combination of coins from the different stacks would allow you to determine which of the three stacks was the light stack in a single weighing. Once the three-stack problem is solved, the same procedure will immediately generalize to the 10-stack problem.

The three-stack problem can be solved in a single weighing only by including some number of coins from each stack and using the amount of underweight as measured by the pointer scale to determine which of the stacks is light. To make use of the information concerning the number of grams by which the weighing is underweight (from what it would be if the coins were all true quarters), we clearly need to have some way of associating the amount of underweight with each of the three stacks. This type of reasoning is an illustration of drawing inferences from the goal (determining the light stack in a single weighing). Again stop reading and try to solve the problem, if you have not done so already.

The procedure required to associate each stack with an amount of underweight is to take one coin from the first stack, two from the second stack, and three from the third. If the pointer scale reads 2 grams underweight, you know that the first stack is light. If it reads 4 grams underweight, the second stack is light. If it reads 6 grams underweight, the third stack is light. Generalize the solution to the 10-stack problem, if you have not done so already.

The original 10-stack problem is solved in a single weighing as follows: You take one coin from stack 1, two coins from stack 2, three coins from stack 3, and so on, up to 10 coins from stack 10. Now weigh this entire set of coins and determine by how many grams it is underweight. The number of grams of underweight divided by 2 is the number of the stack that is light. Thus, the solution can be achieved with only a single weighing, when you have a sufficiently large number of coins available in each stack. Reducing the number of coins in each stack does not simplify the problem; indeed, it makes the problem much more difficult, to the point of completely preventing you from seeing the elegant solution to the original problem. When you reduce the number of *stacks* rather than the number of *coins* in each stack, you obtain problems that are in some sense simpler, though, of course, you cannot reduce the number of weighings below 1.

Reducing the number of elements in a problem is not the only way to make it simpler. Another way is to change the problem so as to allow you to use an already proved theorem in the solution. That is, you change the problem so that it permits you to use a theorem or knowledge that you do not already know how to use in the original problem. You then hope that applying this theorem to the simpler problem will give you an idea of how to use the theorem in the original problem. A good example of this technique is provided by the following distance problem:

> A ray of light travels from point A to point B in Fig. 9-4 by bouncing off a mirror represented by the line CD. Determine the point X on the mirror such that the distance traveled from point A to point B is a minimum. What is the relationship between the angles α and β?

Stop reading and try to solve this problem.

Probably the most salient piece of knowledge we all have about minimum distance is the geometric assumption that the shortest distance between two points is the straight line connecting them. However, it is not immediately apparent how to apply this knowledge in

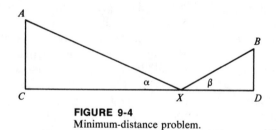

FIGURE 9-4
Minimum-distance problem.

the present problem, since we are constrained to connect A and B with a bent line that touches the line CD at some point X. If you have not solved the problem, stop reading and try to define a simpler problem that allows you to apply the principle that the shortest distance between two points is the straight line connecting them.

The difficulty in applying this principle is that the points A and B lie on the *same* side of the line CD (which the shortest-distance line is required to intersect). This fact prevents the shortest line from being a straight line. However, if the points A and B lay on *opposite* sides of the line CD, it would then be possible to intersect the line CD with a straight line connecting the points A and B. This statement suggests the definition of a simpler problem in which the points A and B lie on opposite sides of the line CD. Try to define such a problem, if you have not done so already.

A simpler problem that permits the application of the shortest distance principle is the following:

> Find the point X such that the distance from point A to point E passing through the line CD in Fig. 9-5 is a minimum.

Since points A and E lie on opposite sides of the line CD in this new problem, the shortest distance between points A and E will be the straight line connecting them. The point X will be the intersection of line AE with line CD. Now if point E is constructed as indicated in Fig. 9-5 to be the same distance from the line CD, and at a point symmetrically opposite point B, then it is obvious that the distance XB will be equal to the distance XE (since these are corresponding parts of congruent triangles). This indicates that the solution for point X in the simpler problem is actually the solution for point X in the original problem. Furthermore, $\alpha = \gamma$, since they are opposite interior angles of intersecting straight lines, and $\beta = \gamma$, since they are corresponding angles of congruent triangles. Thus, $\alpha = \beta$ (the angle of incidence equals the angle of reflection), and the original problem is entirely solved.

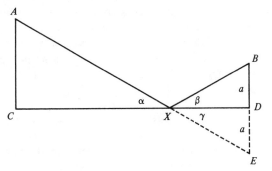

FIGURE 9-5
Simpler minimum-distance problem.

Now that you have had some experience with minimum-distance problems, perhaps you would like to try your hand at my version of the classic *walking-fly problem*:

> Billy Smith smudged his lollypop at a point on the wall of the living room 1 foot from the floor and 6 feet from each corner. A fly with a broken wing is standing on the opposite side wall 1 foot from the ceiling and 6 feet from each corner. If the living room is 30 feet long, 12 feet high, and 12 feet wide, what is the shortest path along which the fly should walk to get from where he is to the lollypop smudge?

Stop reading and try to solve the problem.

Since this is obviously a minimum-distance problem, it is similar in that respect to the preceding problem and thus perhaps may be approached in a similar way. In deciding how to apply the methods used in a previous problem to a new problem, it is important to state what you did in the preceding problem at some level of abstraction that is general enough to apply to both problems. In trying to state such an appropriate level of generality, you may begin by stating what you did in whatever manner comes to mind most quickly. This statement may well be too specific to apply to the present problem, but you might then try to state your methods in progressively more and more abstract form, until you reach some statement that applies to the present problem. Now stop reading and try to formulate (at perhaps several levels of generality) what was done in the preceding minimum-distance problem, in order to get ideas for the present minimum-distance problem.

One thing that was done in the preceding problem was to reflect a point about a line in order to construct an equivalent distance for which the solution was a straight line. You might investigate the possibility of reflecting the starting point, goal point, or other points along the walls, floor, and ceiling of the room in the walking-fly problem, but this procedure will not produce a solution. Thus, although reflection about an axis is an operation that can be performed in the walking-fly problem, this operation will not help solve the problem. Can you think of a more general way to state what was done in the preceding problem that may suggest other operations to apply to the walking-fly problem and produce a solution? Stop reading and try to solve the problem, if you have not done so already.

A simple way to make something more general is to strip it of some of its properties, leaving these properties unspecified. To solve the preceding minimum-distance problem, we performed some operations so as to construct an equivalent problem for which the solution is a straight line. Now stop reading, and try to perform some operation such that you obtain an equivalent problem to which the solution is a straight line, if you have not solved the walking-fly problem already.

The only way the fly could follow a straight line in the original room is to fly across the room, which he cannot do because of his broken wing. Thus, we must construct a new medium through which the fly can walk an equivalent distance in a straight line from the starting point to the goal. The reflecting of a point in the previous minimum-distance problem could also be considered to be a rotation of a strip of paper containing the point 180° around the line axis shown in Fig. 9-5. In order to rotate only one and not both points, it would obviously be necessary to cut the paper at an angle along a perpendicular to the axis of rotation. Can you think of some way of cutting up the room and rotating some combination of walls, floor, and ceiling that would result in a completely flat two-dimensional surface? Having achieved an equivalent flat surface, you could then connect the starting and finishing points by a straight line. Stop reading and try to solve the problem, using this hint, if you have not done so already.

As a child, you may have made boxes out of flat pieces of paper. Since the living room in the fly problem is equivalent to a rectangular box, it is just as possible to cut along various edges and flatten it out as it is to construct it from an originally flat piece of paper. Thus, we can obtain a flattened analog of the living room as shown in Fig. 9-6.

Having flattened out the room in the manner shown in Fig. 9-6, it is a simple matter to determine that a straight line connecting the given and the goal is the hypotenuse of a right triangle whose sides are 24 feet and 32 feet. Thus, using the Pythagorean Theorem, we find the length of the hypotenuse equals $\sqrt{24^2 + 32^2} = \sqrt{576 + 1024} = \sqrt{1600} = 40$ feet. Thus, the fly must travel a distance of 40 feet, and the path

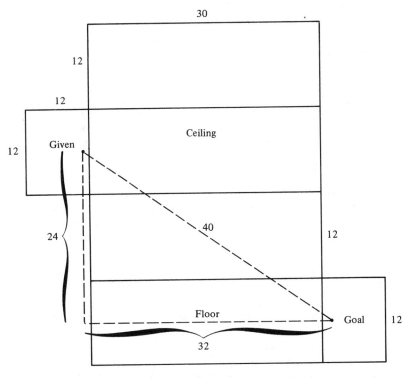

FIGURE 9-6
Flattened living room for the walking-fly problem.

he must follow involves traveling across his own end wall, a portion of the ceiling, a portion of one of the long side walls, a portion of the floor, and a portion of the opposite end wall.

More Complex Problems

Posing a problem that is more complex than the given problem is the logical inverse of posing a problem that is simpler than the given problem. Offhand, it would seem that posing a more complex problem would hardly be a useful problem-solving technique, and in general this is true. However, if all else fails, you might attempt to pose a more complex problem in which the elements of the given problem are included with additional complications, just on the chance that it would give you some ideas. This method is clearly a last resort and unlikely to be beneficial, but it is worth considering for one reason—namely, you may already have solved a more complex problem in which your present problem is essentially embedded. If this is the case, then thinking of a more complex related problem that you have already solved will provide you with all the ideas necessary for the solution of your present, simpler problem. I do not know of many examples of this, but here is one:

> Given a five-by-five checkerboard, as shown in Fig. 9-7, try to draw a line through all the squares of the checkerboard, starting from the square with the dot in it on the left side and passing through each box once and only once, without ever lifting pencil from paper and without ever passing outside of the checkerboard. Show how to do this or prove it impossible.

Stop reading and try to solve the problem by considering a previously solved related problem.

One problem that certainly has some relation is the notched-checkerboard problem discussed in Chapter 3. If you recall how the notched-checkerboard problem was solved, it might well provide you with all the ideas needed to solve the present problem. Now stop reading and try again to solve the problem, if you have not done so already.

The problem that is most closely similar is the integer-path addition problem discussed in Chapter 7. There the problem was to place the integers 1 to 9 in a continuous path over a three-by-three matrix such that the three-digit number in the first row plus the three-digit number in the second row summed to the three-digit number in the third row. Essentially the integer-path addition problem involves drawing a line starting from one cell of a three-by-three matrix (checkerboard) in

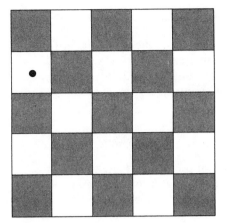

FIGURE 9-7
The five-by-five checkerboard.

precisely the same way as is involved in the present problem using a
five-by-five matrix (checkerboard). The integer-path addition problem
also involved another restriction, making it a more complex problem
than the present one. However, the solution to the integer-path addition problem involved a consideration of the restriction on possible
solutions placed by the path (continuous line) aspect of the problem.
Now stop reading and try to solve the problem, if you have not done
so already.

In both the notched-checkerboard and the integer-path addition
problems the critical property is the imposition of a checkerboard
coloring pattern on the five-by-five checkerboard. Now note that every
time you draw a line through two squares, you necessarily draw a line
through one white and one black square. Alternatively, you could impose a two-dimensional coordinate labeling scheme from (1, 1) to
(5, 5). In that case, notice that, whenever you leave a square with an
odd coordinate sum, you pass into a square with an even coordinate
sum, and whenever you leave a square with an even coordinate sum,
you pass into a square with an odd coordinate sum. Now stop reading
and try again to solve the problem, if you did not before.

Consideration of the implications of checkerboard coloring patterns
for the present problem yields the following inference: If you start
in a white square and must draw a line through an odd number of
squares in total, then the color of the last square you pass through must
be the same color as the square you started from. In the present instance, there are 25 squares in the five-by-five checkerboard. Thus,

if you begin in a white square, you must end in a white square, and there must be exactly 13 white squares and 12 black squares in the checkerboard. However, in the checkerboard shown in Fig. 9-7, there are 13 black squares and 12 white squares. Thus, starting from any white square on the board, it will be impossible to solve the problem of drawing a continuous line through each square once and only once.

SPECIAL CASE

Particularly in proof problems, it often happens that the theorem to be proved states a general relation that holds over a number of special cases or entities. In such problems, it is often useful to try to prove the theorem first for one or more of these special cases before an attempt is made to prove the theorem in general. The reason is that it is usually easier to prove the theorem for a special case than for the theorem in general. This argument is precisely the same one made for the advantages of posing and solving simpler problems in general. However, not all simpler problems are special cases of the problem you are trying to solve. The reverse, however, is almost invariably the case — special cases are simpler problems than the general problem.

Proving a theorem true for one or more special cases increases the probability that the theorem is true in general, but unless you can prove the theorem true for all special cases, proving the theorem in a particular case does not, of course, prove the theorem in general.

However, disproving a special case of a conjectured theorem does disprove the theorem in general. When you are uncertain about the truth of the theorem, it can be particularly useful to investigate the theorem in some special case, since a quick disproof of the theorem for the special case disproves the theorem in general. This exercise may save you considerable time that otherwise might be spent in fruitless attempts to prove a false theorem.

When the theorem is true, proving it true for one or more special cases may provide you with many of the elements needed in order to prove the theorem in general. This reason is perhaps the primary one for posing and solving special cases of general problems.

One use of the method of special case was discussed already in Chapter 6 on subgoals as a part of the method of mathematical induction. Recall that, in the method of mathematical induction, we had to first prove the theorem true for $n = 1$ (a special case) and then show that if the theorem was true for n it was true for $n + 1$. Thus, in proving that the sum of the first n integers equals $n(n + 1)/2$, we initially established that this was true for $n = 1$.

Another use of the method of special case occurs sometimes in multiple-choice examination questions. For example, if you were asked to choose one of five formulas for the sum of the first n integers, the fastest method might be to investigate each formula on some special case, such as $n = 5$, very likely determining that all but one of the answers produced a contradiction in that special case. Note that this is, in essence, a combination of the use of two problem-solving methods, namely, special case and the method of contradiction.

A similar problem often arises when you try to remember some formula you learned previously and think you recall it but are not sure. For example, in trying to recall the formula for the sum of the first n integers, you might erroneously recall something such as $n(n - 1)/2$. Such erroneous conjectures can easily be tested and rejected by investigating their truth in one or more special cases. Since you often have a reasonably good idea of what the correct formula is, a few rejections of incorrect statements of the formula will usually be followed by a correct statement, which might simply be verified by mathematical induction.

Deriving a formula for the number of combinations of m things taken n at a time provides another good example of the use of the method of special case. Undoubtedly you the reader have encountered this formula in the past; however, in my experience, many students fail to remember the formula and most do not know how to derive it. Even if you do know how to solve the problem, it is useful to think of how you would go about applying the method of special case to derive the formula. Thus, consider the following:

Derive a formula for the number of combinations of m things taken n at a time $(m > n)$. Combinations refer to the number of different *unordered* sets of elements. That is, the set of two elements obtained by drawing X and then drawing Y is equivalent to the set obtained by drawing Y and then X. The set XYZ is equivalent to the set YXZ or the set ZYX. The ordering of the elements in the set is irrelevant. Furthermore, you are restricted to drawing an element only once from the underlying population of m elements. That is, you may sample from the underlying population n times without replacing the elements you sampled (sampling without replacement).

Stop reading and try to solve this problem, making use of the method of special case.

There are four specific aspects to the problem. First, there is an underlying population of m elements. Second, you are picking a sample of n of these elements. Third, the sampling is done without replacement; that is, every time you pick an element from the sample, you do not put it back in the population, so the population is reduced by one element every time you choose an element for the sample. Fourth, you are concerned with the number of different unordered sets obtained by this sampling procedure, rather than the number of different ordered sets.

Each of these four aspects could be changed to pose a problem related to the present one, some of them simpler than the present problem, which might facilitate its solution. If you have not yet solved the problem, stop reading and think how you might change one or more of the four aspects to derive a related problem or a special case that is simpler to solve than the original problem.

You could reduce the size of the underlying population of m elements to, say, the special case of two elements. You would then also have to reduce the size of the sample to either one or two elements ($n = 1$ or 2). However, it is probably unnecessary to reduce both m and n in this way. It is quite possible to leave m as it is and reduce the sample size to two elements ($n = 2$). You would now be considering the special case of the present problem where $n = 2$, namely, where one is selecting an unordered pair without replacement from the underlying population of m elements. Stop reading and try to solve the problem, if you have not done so already.

It will be easier to solve for the special case of the number of unordered pairs of elements if you consider the related problem of determining the number of ordered pairs of elements that can be selected from the population of m elements. This latter problem is quite trivial to solve: there are m ways to select the first element, and for each of these m ways, there are $(m - 1)$ ways to select the second element; thus, there are $m(m - 1)$ ways to select an ordered pair of elements. Having determined the number of ordered pairs, it is quite possible to determine the number of unordered pairs. Stop reading and consider how you would do this, then generalize your answer to solve the original problem, if you have not done so already.

The critical difference between ordered and unordered pairs is that a pair of ordered elements XY is considered equivalent to the ordered pair YX, when unordered pairs are being considered. Thus, there are exactly two different ordered pairs of elements for each unordered pair. Knowing this, how can you solve the problem of determining the number of unordered pairs, if you know how many ordered pair there are? Stop reading and solve this problem and then generalize your answer to solve the original problem, if you have not done so already.

Clearly, if there are two ordered pairs for each unordered pair and $m(m - 1)$ ordered pairs, then there are $m(m - 1)/2$ unordered pairs.

Now generalize this answer to the solution of the original problem, where you are selecting not a pair of elements but a set of n elements at a time. The relevant generalization of the solution to the special case is that the problem should be broken into two parts. First, the number of ordered sets of elements should be determined, then how many different orderings there are for each unordered set. Now use this analysis of the two subproblems to solve the original problem, if you have not done so already.

If you are not able immediately to generalize the solution of the special case to obtain a solution to the general case, then consider another special case, where $n = 3$. Here the number of ordered sets obtained by sampling without replacement is $m(m - 1)(m - 2)$. The number of different orderings of each sample of three elements is the number of permutations of three things. The number of permutations of a set of three things is $3 \cdot 2 \cdot 1$, or 6, since there are three ways to pick the first element from the sample, two ways to pick the second, and one way to pick the third. Thus, the number of different unordered sets of three elements equals $m(m - 1)(m - 2)/3 \cdot 2 \cdot 1$ $= m(m - 1)(m - 2)/3!$. Stop reading and generalize the formula to the case of an unordered set of n elements, if you have not done so already.

Clearly, the general formula for the number of unordered sets of n elements selected from a population of m elements without replacement is $m(m - 1) \ldots (m - n + 1)/n! = m!/n!(m - n)!$. The general principle for solving this problem comes from breaking it into two parts in order to determine, first, the number of ordered sets and, second, the number of distinct ways of ordering (reordering) the elements in a particular ordered set (sample). The number of unordered sets is equal to the number of ordered sets divided by the number of ways of ordering the elements in a particular set. This general principle is essentially present in all special cases where $n = 2, 3$, and so on. Thus, solving the problem for one or two special cases provides all the essential ingredients for solving the problem in general.

The method of special case is also frequently useful in geometric problems. Consider the following proof problem in Euclidean geometry:

You are given the following: (a) A straight line equals an angle of $180°$. (b) A right angle equals $90°$. (c) If two parallel lines are cut by a transversal, the alternate interior angles are equal. Prove that the sum of the angles of any triangle equals $180°$.

Stop reading and try to prove this theorem, making use of the method of special case.

Since you are given information on the number of degrees in the right angle, it is reasonable to consider the special case where one of the angles of the triangle is a right angle. For example, consider the right triangle shown in the left in Fig. 9-8.

Stop reading and try to solve the problem for this special case, then generalize your answer to a triangle without any right angle, if you have not done so already.

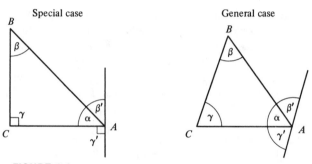

FIGURE 9-8
Proof of the theorem in Euclidean geometry that the sum of the angles of a triangle is 180°. Left is for the special case of a right triangle. Right is for the general case of any triangle.

Given the right triangle shown in Fig. 9-8, it would be reasonable to construct a line at point A parallel to the opposite side of the triangle. Having done this, the alternate angles are equal: $\beta' = \beta$ and $\gamma' = \gamma$. Since $\gamma' = \alpha + \beta'$, then $\alpha + \beta' = 90°$; this establishes that $\alpha + \beta = 90°$. Thus, $\alpha + \beta + \gamma = 180°$, and the theorem is proved for the special case of a right triangle. Now stop reading and try to generalize your solution to the case of any triangle, if you have not already solved the problem.

In solving the special case, we have essentially mapped the angles of the triangle onto three angles that, taken together, form a straight line. This approach extends in a direct way to the general case for any triangle, not just a right triangle. Thus, for the general triangle, we would be led to construct a line at A parallel to the opposite side. Then, we would go through exactly the same reasoning as in the special case. We might not recognize in the special case the general principle of mapping the angles of the triangle onto a straight line. However, it is likely we would consider erecting a line at A parallel to the opposite side of the triangle in the special case, noticing that $\alpha + \beta' = 90°$ and that $\beta = \beta'$. This procedure includes the essential notion of constructing an additional line at A parallel to the opposite side, and thus, you might think of the more general principle.

In some instances, it turns out that proving the theorem for a small number of special cases constitutes the proof of the entire problem. Sometimes this result is obvious in advance, and sometimes it only becomes obvious after considering a special case. For example, in Chapter 7 we considered the problem of proving that $A^2 - AB + B^2 > 0$. We considered the proof of this theorem only for the case where $A > 0$ and $B > 0$ but noted that the theorem was actually true for all A and B. Proving the theorem for all A and B essentially requires us to prove it for four cases: (a) $A > 0$ and $B > 0$; (b) $A > 0$ and $B < 0$; (c) $A < 0$ and $B > 0$; (d) $A < 0$ and $B < 0$. The proof of the theorem for each of these four special cases involves breaking the problem up into three more special cases within each of the four previously mentioned special cases — namely, $A > B$, $A = B$, and $A < B$. Thus, in all, we have 12 special cases for which to prove that $A^2 - AB + B^2 > 0$ is true. But for each of these 12 special cases, the theorem is rather simple to prove.

An example in geometry where the general problem can be divided into two essentially identical special cases is provided by the proof of the following theorem:

> You are given the following: (a) The measure of an intercepted arc in degrees is the same as the measure of its corresponding central angle (namely, the angle determined by drawing the radii from the center of the circle to the ends of the intercepted arc). (b) The sum of the angles of a triangle equals $180°$. (c) The angles opposite the equal sides of an isosceles triangle are equal. Prove that an angle inscribed in a circle has half as many degrees as its intercepted arc.

Stop reading and try to solve the problem by first considering a special case.

As is frequently true, there are many ways of formulating special cases of the present theorem. For example, we might consider the special case where the inscribed angle is a right angle and try to prove that its intercepted arc is $180°$ (that the cord for this arc is a diameter of a circle). Conversely, we might investigate the special case in which the intercepted arc was $180°$ and try to prove that the inscribed angle was $90°$. Another type of special case would be to assume that the cords composing the inscribed angle were equal, and so on. We might consider many special cases before we hit on a special case that is most useful in solving the general problem. No previously mentioned special case is optimal for the solution of the present problem, but the alternate special cases may suggest such a one. Stop reading and try to solve the problem, if you have not done so already, by using the method of special case.

The optimal special case to consider is that where one of the sides of the inscribed angle is a diameter of the circle. This special case is illustrated in the upper section of Fig. 9-9. Now stop reading and try to solve the special case, then extend your answer to a proof of the general theorem, if you have not proved the theorem already.

Proving the theorem for the special case is relatively straightforward. First, draw in the dashed line shown in the upper circle of Fig. 9-9, to obtain the central angle β, which we know is equal to the intercepted arc. We can easily verify that the triangle shown in the figure is isosceles. Now making use of the assumption that the angles opposite equal sides of an isosceles triangle are equal, we know that

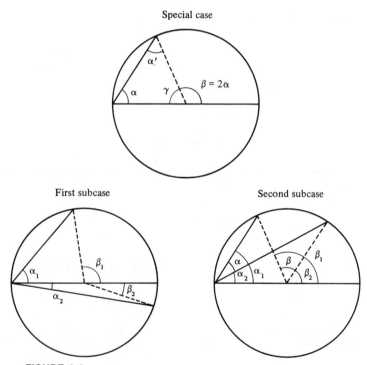

FIGURE 9-9
Diagrams for the proof of the theorem in Euclidean geometry that an inscribed angle has half as many degrees as the intercepted arc (which equals the central angle).

the inscribed angle $\alpha = \alpha'$. Therefore, by the givens that the sum of the angles of a triangle equals 180°, we know that $\alpha + \alpha' + \gamma = 180°$. Therefore, $2\alpha = \beta$ or $\alpha = \beta/2$, and the theorem is proved for the special case. Now stop reading and extend your solution to the general case. Actually, it is probably simpler to consider two types of general cases. In essence, we are subdividing the general case into two special cases that exhaust the entire category. Let us call these two cases the first subcase and the second subcase. In the first subcase, the inscribed angle includes the diameter of the circle drawn from the vertex of the angle. In the second subcase, the inscribed angle does not include the diameter drawn from the vertex of the inscribed triangle. Now stop reading and try to solve the problem for each of the two subcases of the general case, if you have not done so already.

If in the first subcase we can divide the angle α into two components, then $\alpha_1 + \alpha_2 = \alpha$ such that the dividing line for the two component angles is a diameter of the circle. Now for each of these component angles, $\alpha_1 = \beta_1/2$ and $\alpha_2 = \beta_2/2$, as shown in the lower left diagram of Fig. 9-9. Thus, $\alpha_1 + \alpha_2 = (\beta_1 + \beta_2)/2$, and the theorem is proved for the first subcase. Now stop reading and solve for the second subcase, if you have not already done so.

In the second subcase, we can consider the inscribed angle α to be equal to the difference between angle α_1 and angle α_2, as illustrated in the right-hand diagram of Fig. 9-9 ($\alpha = \alpha_1 - \alpha_2$). Since each of the component angles, α_1 and α_2, satisfies the requirement of the special case (that one of the cords of the angle be a diameter of the circle), we know that $\alpha_1 = \beta_1/2$ and $\alpha_2 = \beta_2/2$. Thus, $\alpha = \alpha_1 - \alpha_2 = \beta_1/2 - \beta_2/2 = (\beta_1 - \beta_2)/2 = \beta/2$, and the theorem is proved for the second subcase.

This problem provides a beautiful example of the multiple use of the method of special case. An extremely special case was first investigated to get the basic idea for the solution. Then the general case was subdivided into two special subcases, which were nevertheless more general than the original special case. In proving the theorem for each of these two more general special cases, the truth of the theorem for the special case was used as an integral part of the proof.

In some cases, the solution of a single special case may provide the solution to the general problem. One example of this is provided by the following problem:

A cylindrical hole 10 inches long is drilled through the center of a solid sphere, as shown in Fig. 9-10. What volume remains in the sphere?

Stop reading and try to solve the problem, using the method of special case.

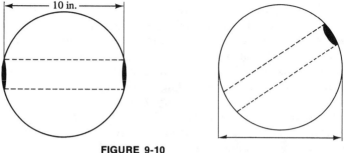

FIGURE 9-10
The hole in the sphere problem.

The problem implies that the volume remaining in the sphere is independent of the diameter of the cylindrical hole, provided that the hole is 10 inches long. Assuming that the problem has a unique solution, independent of the diameter of the hole, we can get a solution for the general problem very simply by considering a special case. What is this special case? Stop reading and try to solve the problem, if you have not done so already.

Consider the special case where the cylindrical hole has a diameter of zero. In this case, since the cylindrical hole is 10 inches long, the sphere must have a diameter of 10 inches, and the volume of a solid sphere with a diameter of 10 inches equals $\frac{4}{3}\pi r^3 = \frac{4}{3}\pi 5^3 = 500\pi/3$. Making the assumption (which we have certainly not proved) that this problem has a unique solution, independent of the width of the hole, $500\pi/3$ must be the answer. Naturally, if we were conjecturing that the validity of this theorem was uncertain, we could not use this line of reasoning to solve the problem.

A second example of a general problem that can be solved by solving a single special case is the following:

> In this two-person game the players alternately place poker chips on a circular table. The chips must not overlap and must be completely on the table; that is, no poker chip may stick out over the edge of the table. The last player to play a chip on the table is the winner. If each player makes the optimal move on his turn, will the first player or the second player be the winner?

Stop reading and try to solve the problem by considering a special case.

The problem suggests that optimal strategy will produce a forced win for either player 1 or player 2, independent of the size of the table. Assuming this, what special case yields a quick solution?

Consider a table that is big enough to accommodate only one poker chip (when placed in the center of the table). In such a case, the player who goes first will be able to place the first and the last poker chip on the table and will therefore be the winner. This case suggests that, if one of the players has a forced win playing by optimal strategy for all sizes of tables, then that player is the first player. Verifying this hypothesis for a table of any size requires a further clever insight beyond that provided by the special case. However, the special case does suggest that we should test the hypothesis that it is the first player who can force a win for himself by optimal play. Also, the winning first move for the first player in the special case might suggest the winning first move for the first player in the general case. Stop reading and try to solve the general case, if you have not done so already.

The insight involved in solving the general problem is that the first player initially places a poker chip in the center of the table (as in the special case) and thereafter plays chips in a symmetrically opposite position to that played by the second player. Clearly, if the second player has any place on the table available to place a poker chip, there will still then be a symmetrically opposite place on the table for the first player to place a chip, so that the first player must be the last to play a chip on the table, independent of the size of the table. Note that the initial, unique move by the first player—namely, placing a poker chip in the center of the table—is exactly the same move that the first player should make in the special case.

A third example of a general problem that can be solved by solving a single special case is the following:

Triangle *ABC* is formed by three tangents to a circle, as shown in Fig. 9-11. Angle *DAE* = 26°. Solve for angle *COB*.

Stop reading and try to solve the problem.

Angle *DAE* (which is the same as angle *BAC*) is completely determined by two of the three tangents. It is impossible from the information given in the problem to determine the location of the tangent *BC* as it intersects the circle anywhere within the arc *ED*. Thus, assuming that the problem has a unique solution for angle *COB* (which is certainly implied by the statement of the problem), we can solve for angle *COB* by considering any special case of the placement of tangent *CB*. Stop reading and try to solve the problem, if you have not done so already.

The obvious special case to consider is for tangent *CB* to intersect the circle at the same point as does the line from the origin of the circle to point *A* (as shown above in Fig. 9-11). Having chosen this

178

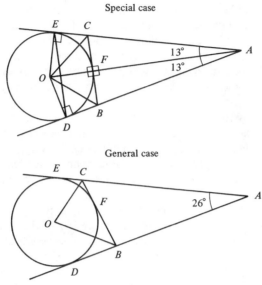

FIGURE 9-11
Diagrams for solution of the problem to find the angle COB.

special case, it is now relatively easy to solve the problem. Since angle AED and angle ADE intersect the same arc of the circle, these two angles are equal. Thus, $AE = AD$, since the sides opposite equal angles of a triangle are equal. Line $OE =$ line OD, since both are radii of the same circle. Line $OA =$ line OA, since they are the same line. Thus, triangle AOE is congruent to triangle AOD, by virtue of having the three corresponding sides equal. Thus, angle $EAO =$ angle $DAO = 13°$, since angle $DAE = 26°$. Since there are 180° in a triangle, and angle $AEO =$ angle $ADO = 90°$, we know that angle $AOE =$ angle $AOD = 90° - 13° = 77°$. We can easily prove that triangle OEC is congruent to triangle OFC by having the same hypotenuse and one equal side (radii of the circle). Similarly, triangle ODB is congruent to triangle OFB. Thus, angle $COF = \frac{1}{2}$ angle EOF and angle $FOB = \frac{1}{2}$ angle FOD. Putting all this together implies that angle $COB = 77°$, and the problem is solved.

A fourth example of solution of a general problem by means of a single special case is the following:

In Fig. 9-12 angle $BAD = 20°$, $AB = AC$, and $AD = AE$. Solve for angle CDE.

Stop reading and try to solve the problem, making use of the method of special case.
This problem first seems to lack enough data to solve it. The given information in the problem is not adequate to specify a single unique triangle with these properties. There are a large variety of different (noncongruent) triangles consistent with the information that $AB = AC$, $AD = AE$, and angle $BAD = 20°$. Furthermore, these triangles are not even similar to one another; that is, the angle DAE can assume a variety of different values. Clearly, the absolute lengths of the sides AB, AC, AE, and AD are not relevant to determining the angle CDE.

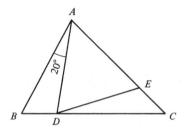

FIGURE 9-12
Geometry problem.

However, it is surprising to be told implicitly by the problem that the angle DAE is irrelevant to the value of the angle CDE. Assuming that within some range of values the magnitude of angle DAE is irrelevant to the magnitude of the angle CDE, then how might we go about solving for the magnitude of angle CDE? Stop reading and try to solve the problem, if you have not done so already.

Clearly, we can solve the problem for a special case of the value of angle DAE and determine the magnitude of the angle CDE. According to the implicit information stated in the problem, we should obtain the same solution for the angle CDE irrespective of our choice of angle DAE (over some range). Thus, let us pick angle $DAE = 20°$. Now stop reading and solve for angle CDE in this special case, if you have not already solved the problem.

If angle $DAE = 20°$, then angle $BAC = 40°$ and angle $ABD = ACD = 70°$, since these angles are opposite the equal sides of an isosceles triangle and there are 180° in a triangle. By the same reasoning, angle $ADE = $ angle $AED = 80°$. Thus, angle $DEC = 100°$, and therefore angle $CDE = 180° - 100° - 70° = 10°$. It turns out that an infinity of other values substituted for the angle DAE yield the same value (10°) for angle CDE, which is the solution to the problem. In general, we could simply substitute some arbitrary value y for angle DAE and solve the set of equations to determine the value of angle CDE in a manner that would be independent of y. However, it is considerably simpler to solve the problem by choosing a single special case, since the given information implies that the solution to the special case is equivalent to the solution to the general problem.

GENERALIZATION

Just as it is sometimes useful to solve a special case prior to working on a more general problem, it is also frequently useful to do the opposite and generalize the problem somewhat.

Generalization plays a role in problem solving in at least three different ways. First, as a necessary part of problem solving, we usually abstract from a problem certain properties belonging to a more general class of problems and thus relevant for determining the previously established principles needed for solving our present problem. Second, after we have solved the problem, it is often useful to consider whether we could generalize a solution from it to a wider class of problems in order to derive a more general conclusion or one or more corollaries of the principle established in the problem we just solved. Third, occasionally (though in my experience not too frequently), it may be useful to pose and attempt to solve a more general problem prior to working on the current problem, even when the solution to that more general problem is not yet known to us.

The first role that generalization plays in problem solving has really already been discussed in Chapter 3 in connection with the representation of information. Recall that a critical aspect of solving many problems consists in retrieving from memory the relevant previously established relations and principles with common properties needed to solve the present problem.

It may be that the current problem is really a special case of a general class of problems for which we already know a simple rule for solution. For example, if the present problem is the linear equation $2x + 5 = 13$, we know that the solution to this particular linear equation can be achieved by using the general methods for solving any linear equation of the form $ax + b = c$. Similarly, if the equation were a quadratic of the form $7x^2 + 2x - 4 = 0$, we have a formula for solving any equation of the form $ax^2 + bx + c = 0$. A broad range of higher order equations can be solved by certain types of numerical methods. If we have a particular equation that falls within the scope of a numerical method, we know we can apply this method to solve the particular problem.

In a geometric context, if a problem gives two sides of a right triangle and we are asked to solve for the third side, we know a general method that is applicable to solving all such problems—namely, use of the Pythagorean Theorem, $c^2 = a^2 + b^2$.

If we are given a problem in which we must determine the number of combinations of seven things taken four at a time, we need only retrieve the formula for the number of combinations of m things taken n at a time and substitute in the appropriate values for m and n in order to solve the problem.

Ordinarily, to solve problems we must combine use of more than one previously established principle. Thus, in all proof problems, whether algebraic, geometric, or logical, the proof invariably requires a sequential application of several previously established principles. In a story-algebra problem, first the methods of representing the information algebraically must be applied, then the methods appropriate for solving whatever algebraic equations are derived from the story.

The examples of generalization in this most important context could be extended indefinitely. Suffice it to say that, in this sense, generalization plays an enormously important role in problem solving. However, as discussed in Chapter 3, this use of the method of generalization depends critically on the degree of understanding you have of the previously established principles in the areas relevant to the current problem. A few general principles regarding representation of information are discussed in Chapter 3 and Chapter 10; however, the field is just too vast and thus is outside the scope of this book.

The second role of generalization in problem solving has little use for a student in a course but is frequently valuable for a mathematician or scientist solving a new problem to see if the solution can be generalized to a larger class of problems. Along the same line, we may try to derive some additional consequences as relatively straightforward corollaries of the solution to the present problem. For example, if we prove that the two diagonals of a rhombus are perpendicular, we might then notice that the same property holds for a kite. We need only to have a quadrilateral with sides a and b adjacent, sides c and d adjacent, and $a = b$ and $c = d$. All four sides need not be equal, nor do opposite sides need to be parallel.

Another example in a geometric context is provided by the theorem that the alternate interior angles formed by a transversal intersecting two parallel lines are equal. This result easily generalizes to establish that the other pair of alternate interior angles are equal and that both pairs of corresponding exterior angles are equal as well.

A final example, again in a geometric context: Assume that we have already established that an inscribed angle has half as many degrees as its intercepted arc. From this result it is relatively trivial to show that the angle formed by a tangent and a chord meeting it at the point of contact has also half as many degrees as its intercepted arc. In fact, the latter theorem could be thought of as simply being a limiting case of the former theorem.

The third possible role of generalization in problem solving is, in a sense, the inverse of the previously described role of the method of special case; namely, it might facilitate solution of a specific problem to formulate a more general problem that had not been previously solved. Then we might either solve the more general problem or, in any event, work on the solution of the more general problem for a time, before going back to working on the specific problem.

Polya (1957, pp. 108–109) argues that this problem-solving technique is quite useful and he gives the following example:

> The problem is to find a plane that passes through a given line and bisects the volume of a given octahedron.

Polya asserts that it would be useful to formulate the more general problem of finding a plane that passes through a straight line and bisects the volume of any solid with a center of symmetry. The solution to this problem is fairly obvious, namely, a plane determined by the given line and the center of the solid with a center of symmetry. Since an octahedron is a special case of a solid with a center of symmetry, the original, specific problem is solved. Polya asserts that the value in formulating the more general previously unsolved problem is that it can focus the problem solver's attention on the necessary properties in the original problem that must be used in order to solve it. As Polya himself points out, however, the primary function of generalization was in the formulation of the more general problem. If we had generalized the problem in an inappropriate way — that is, using some property that was not in fact central to the solution of the original problem — then the formulation of the general problem would likely have been of no help and might even have been a hindrance in the solution of the original problem.

A two-dimensional analog of the previous example would be to determine the line that passes through a given point and, say, a given square bisecting the area of that square. A generalization of this problem would be to determine a line that passed through a given point and bisected the area of a given plane-figure with a center of symmetry. Again, formulating the more general problem directly suggests the solution — namely, that the line passes through the given point and the center of symmetry of the figure.

Personally, I am somewhat skeptical of the alleged benefits of trying to solve a more general problem, when the solution to the more general problem is not known to the problem solver. However, I am sure that thinking of possible generalizations of the current problem does often aid the problem solver in realizing all of the properties of the problem, some of which may be the critical properties in order to solve the problem. In this sense, the alleged third role of generalization is very similar to the first role of thinking of generalizations of the problem when the solution of the more general problems is already known. It is a question of representing the important properties or principles that are present in a special problem, and, to do so, an abstraction process is involved. Abstracting the properties from a problem is, in essence, generalization. Thus, once again we see that the role of generalization and the role of representation of information (as discussed in Chapters 3 and 10) are very closely linked and perhaps identical.

10

Topics in
Mathematical Representation

As stated in Chapter 2, problems contain information concerning givens, actions, and goals. The first and most basic step in problem solving is to represent this information in either symbolic or diagrammatic form. *Symbolic form* refers to the expression of information in words, letters, numbers, mathematical symbols, symbolic logic notation, and so on. *Diagrammatic form* refers to the expression of information by a collection of points, lines, angles, figures, directed lines (vectors), matrices, plots of functions, graphs, and the like. Often the same information should be represented using a variety of symbolic or diagrammatic notations. In fact, diagrammatic representation is generally labeled; for example, points, lines, and cells in a matrix have symbols attached to them in the diagram. Of course, problems are stated originally in some form, often relying heavily upon verbal language. The first step in solving such a problem is to translate from the representation given explicitly or implicitly in the original statement of the problem to a more adequate representation.

This chapter is concerned with selected topics in the mathematical or precise representation of information in problems. Although precise representation of the information in a problem is the first step to take in trying to solve a problem, I deferred discussing this important topic to this late chapter of the book for two reasons.

First, although some general statements can be made about the representation of information in a large variety of problems, most of the principles of representation are specific to particular problem areas. Effective representation for problems from some area of mathematics, science, or engineering depends upon knowing centuries of conceptual development in the relevant areas of mathematics, science, and engineering. I doubt that mankind will ever develop a general method for determining what are the useful concepts to define in any particular area. Certainly, no such general principles of how to define good concepts are presented in this book. The best I can do is to present those types of concepts and the principles for representing them that have proved the most useful in a wide variety of areas of formal problem solving. This is what is done in the present chapter, without any claim to completeness (which would be preposterous) and with only minimal claim to logical organization of the concepts and the principles of mathematical representation.

Second, although some of the principles of mathematical representation are reasonably simple and can be communicated to even the most minimally prepared student, some of the principles discussed in the latter half of this chapter are concerned with concepts from various areas of mathematics with which some readers will be unfamiliar. I hope that these readers will profit from the sections on sets, relations, operations, mappings, functions, and real-valued functions of a real variable. However, it seemed wisest to put this material near the end of the book so as not to discourage readers with less mathematical sophistication.

The material in the latter portion of this chapter is really a brief, simple discussion of selected mathematical topics, largely modern algebra and combinatorial mathematics. This material is primarily intended for students who have some background in these topics in college, high school, or grade school new math courses. For such students, these sections are intended as review of the relation of certain mathematical concepts to the general methods of problem solving discussed in this book. For students with no background in set theory, modern algebra, and combinatorial mathematics, these sections may be rather hard going and require considerable study. Such students should consult regular mathematics books concerned with these topics, rather than try to master the material on the basis of the rather brief discussion presented here.

The primary basis for selecting the mathematical concepts discussed in this chapter is their applicability to the puzzle-type problems characteristic of recreational mathematics, which constitute the primary example problems in this book. A large subclass of all recreational mathematics problems consists of "insight" problems, where a major difficulty may be to recognize the important concepts for representing the information in the problem.

REPRESENTATION ON PAPER OR IN THE HEAD

This section has a simple message: use pencil and paper extensively when you are trying to solve problems. Of course, the primary representation of information is in your head, but virtually all problems can be solved faster by representing some of the information on paper (or a blackboard or other writing surface) than they can without a written graphic aid. Written representation of information is useful for both verbal symbolic information and visual diagrammatic information. To try to solve problems without using pencil and paper is to subject yourself to an unnecessary handicap. Although an occasional problem may be solved faster purely "in the head," the vast majority of all problems will be more quickly solved by representing information on paper at an early stage in working on the problem. No one can say for sure why this is so, but there are at least four plausible reasons.

First, writing down the components of a problem focuses your attention on the need to give names (symbols, diagrammatic representation) to each of the important concepts in the problem.

Second, it automatically draws your attention to the information stated in the problem as you attempt to represent that information on paper.

Third, as you derive inferences or get to intermediate stages in the solution of the problem, writing aids your memory for these inferences or intermediate stages at later stages in the solution of the problem. After working on a problem for some time, it is easy to forget some of the given information or inferences you drew from the given information, and some of this information may be helpful later. Having this information written down allows you to use rapid visual scanning to jog your memory for prior concepts and facts that might usefully be combined with the concepts and facts to which you are currently paying attention.

Fourth, problems that involve tables or matrices of information are especially difficult to retain as a visual image purely in the mind. Such information is very efficiently represented by means of a table written on paper. For an example of the importance of constructing tables to represent information, see the Smith, Jones, Robinson problem in Chapter 7. Similar conclusions apply to graphs and other figures, which may be difficult to accurately imagine and remember purely mentally, without graphic aids.

Whatever the reason, experience indicates that pencil and paper representation of information is very useful in problem solving. So do not be lazy. Always have pencil and paper ready when you start to work on problems, and make extensive use of them through all stages of problem solving.

DIAGRAMMATIC REPRESENTATION

When a problem in some way involves spatial concepts — points, lines, angles, directions, vectors, surfaces or plane figures, solids, contiguity, connectedness, inside, outside, around — diagrammatic representation may be an extremely useful aid to symbolic representation, whether verbal, logical, or algebraic. Even when the problem does not seem to involve any spatial concepts, it sometimes happens that you can form an analogy between the concepts in the problem and spatial concepts, so that you could draw a diagram that might be of some aid in solving the problem. For example, overlapping circles might be used to represent overlapping sets, points to represent elements of a set, and sets of arrows to represent mappings from one set to another.

Verbal symbolic representation is probably somewhat more important than visual diagrammatic representation in problem solving and in abstract thinking in general. The communication of the givens, operations, and goals of a problem is largely in verbal symbolic terms. Even when we employ diagrams in the solution of problems, they are usually labeled; that is, symbols are attached to the points, lines, and angles. For example, in solving for the lengths of lines or the magnitudes of the angles between lines in geometric figures, we invariably make extensive use of symbols attached to various points, line segments, or angles in the diagram (see Fig. 10-1).

Furthermore, all the spatial information represented by a diagram like Fig. 10-1 can be represented symbolically without having to employ diagrammatic representation. For example, the spatial information represented in Fig. 10-1 can be represented symbolically as follows: lines a, b, and h meet at common vertex B, lines a and d meet at vertex A, lines d, h, and c meet at vertex D, lines b and c meet at

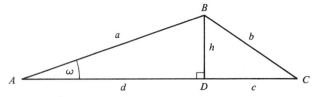

FIGURE 10-1
A diagrammatic representation of the spatial information
in some geometric problem.

vertex *C*, lines *c* and *d* are collinear and line *h* is perpendicular to lines *d* and *c*. If we wished to suppress the symbols for line segments, we could represent lines by unordered pairs of points—for example, (A, B) for line *a*.

By adopting some conventions regarding symbolic representation of spatial or geometric information, the above symbolic representation can be shortened considerably. For example, let the unordered sets of lines meeting at different vertices (points) be represented as follows: (a, b, h), (a, d), (d, h, c), (b, c). The fact that *d* and *c* are collinear (K) and *h* is perpendicular ($\perp$) to them could be represented by something like *d* K *c* and *h* $\perp$ *d* and *c*. So there is nothing uneconomical about the symbolic representation in terms of time to write the information.

However, to say that, compared to symbolic representation, diagrammatic representation is less important is not to say it is unimportant. People could probably learn to solve problems involving spatial concepts with purely symbolic representations such as those just presented, but it is doubtful that they would solve them as efficiently. Current evidence indicates that there is a modality of the mind concerned with spatial concepts that functions differently from the modalities of the mind concerned with verbal symbolic concepts.

The symbolic modalities are much more generally useful (for example, even the spatial concepts can be represented in nonspatial symbolic terms), but it is very likely that the spatial modality of the mind is particularly well suited to reasoning about spatial concepts and problems involving those concepts. Representing the spatial information in a problem in diagrammatic terms probably brings a new part of your mind to work on the problem. Furthermore, that part of your mind is probably very well designed for reasoning regarding the spatial aspects of the problem that are represented in diagrams. Finally, much of your prior knowledge regarding spatial concepts, relations, and so on, is probably stored in the mind's spatial modality. Since such prior knowledge is often assumed implicitly to be part of the givens in a problem, it is clearly important that you have access to your memory for such information.

SYMBOLIC REPRESENTATION

General Principles of Concept Representation

The simplest and most frequent step in symbolic representation of the information in problems is to choose some symbol (or sequence of symbols) to stand for a concept. The concept represented by a symbol can be anything the human mind can conceive. Let us take the symbol x and examine a few of the many concepts it can represent in various problems. The symbol x can represent any real number or it can represent a particular, but (in the present problem state) unknown, real number. Alternatively, x can range over the integers (. . . , -2, -1, 0, 1, 2, . . .), positive integers, negative integers, rationals, irrationals, complex numbers ($y + zi$), the elements of some set, subsets of three elements from some given larger set, and so on. The symbol x can be the label of a particular point, line, or figure in a geometric problem, the label of some particular element in a practical construction problem (such as a gate, a piece of a fence, or a stump), or one of the tokens in a puzzle or game (such as ticktacktoe).

Any symbol can be used to represent any concept, subject to one and only one restriction: the same symbol should almost never be used to represent two concepts that are not known to be equivalent throughout the present problem. There are many other principles of effective representation of concepts, but we can violate them without the risk of producing conclusions that contradict the given information in the problem. We cannot represent by the same symbol two concepts that are not equivalent in a problem, without running a substantial risk of generating contradictions to the given information and deriving incorrect answers.

By contrast, it is perfectly safe to use different symbols for concepts or quantities that may later prove to be equivalent or equal. However, it reduces the load on the memory, if you notice such equivalences or equalities before you assign symbols to concepts or quantities, and assign the same symbol to concepts that must be equivalent. Nevertheless, in view of the grave danger involved in mistakenly using the same symbol for nonequivalent concepts, it is best to always use different symbols for different concepts, unless you are absolutely certain that the concepts are equivalent throughout the problem.

Mnemonics and Symbol Conventions

Although you may use any symbol you like to represent any concept, subject to the above-mentioned restriction, people tend to develop habits regarding the types of symbols used for different types of concepts. For example, in mathematics, i, j, k, l, m, and n tend to be used for variables that range over the integers; u, v, w, x, y, and z tend to be used for variables that range over real numbers; p and q are usually used for probabilities (ranging over the real numbers between 0 and 1); and f, g, and h tend to be used primarily to represent unknown functions.

In scientific and engineering problems, symbols chosen to represent different concepts tend to have some easy mnemonic relation to a longer name for the concept in verbal language, usually being the first letter of the name (or the first letter of the key word, if the name of the concept is some phrase containing several words). For example, f and F might be used for *forces*, A for *area*, p and P for *pressures*, t and T for *times*, r and R for *rates*, and w and W for *work*. Other symbolic naming conventions in science and engineering may be purely arbitrary, such as using Greek letters (θ, α, β) for angles; but adhering to such conventions (whether mnemonic or arbitrary) makes it easy to remember in any particular problem what concepts the symbols represent. Maintaining consistency across different problems in the types of symbols you use to represent types of concepts brings long-term memory to the aid of short-term memory in recalling what is meant by each symbol you are using in any problem.

Single Symbols

If ease of remembering what a symbol represents is so important for efficient problem solving, why not use two or three letters from the full name or even the full name for the concept? Something like this is occasionally done in problems where the names of two or more concepts have the same first letter; for example, using t_o for *total*, t_e for *tension*, and t_i for *time*, when all three concept names appear in some problem. This sort of multisymbol naming of a concept is useful in some cases, and these cases will be discussed in subsequent sections on the use of subscripts, vectors, and functions. However, in cases where the mnemonic advantages of using several symbols to represent a single concept are the only advantages, this practice is almost always a mistake. Unless there are a very large number of concept names that all have the same first letter, it is always possible to think of different single symbols to represent each concept, such that each symbol has an adequate mnemonic relation to the concept name so abbreviated.

For instance, you could use some letter in the name other than the first, capital letters as well as small letters (for example, T and t), corresponding Greek letters (for example, τ for t), phonetically similar letter names if you know phonetics (for example, d for t), or change the concept name to some functional synonym (for this problem) that starts with a different first letter (for example, *sum* for *total*, *force* for *tension*, *duration* for *time*).

For the purposes of unique representation in any one problem, it is unnecessary to represent concepts by a string of several symbols (as we do in verbal language), because so few different concepts are involved in any given problem. You do want to maintain a strong mnemonic relation between your chosen symbol and the concept name it abbreviates, because that provides you with access to the information stored in your long-term memory concerning the concept and its previously established relations to other concepts. However, we have just seen that a single symbol is usually adequate to accomplish this relation without a lot of table lookup.

This being the case, there are substantial cognitive advantages to using single symbols to represent concepts in problem solving. It is fair to say that psychology does not know for sure the exact reasons why this is so, but the experience of problem solvers establishes that it is easier to work with single-symbol names for concepts. Many plausible reasons can be given. Single symbols probably place less of a load on short-term and long-term verbal and visual memory. So, for example, it is easier to remember the visual image or verbal statement of an expression, formula, or equation that uses single symbols for each concept than one which uses more symbols to represent each concept. Single symbols take less time to write on paper and generally have less potential for erroneous writing or reading.

Expressions

To reduce the cognitive and memory load in solving problems, you can also reduce the number of different symbols used to represent concepts by recognizing the relations between concepts right away when you are deciding upon representation. For example, if John is 30 pounds heavier than Bill, let b represent Bill's weight and let $b + 30$ represent John's weight, to avoid assigning any new symbol at all to represent John's weight. This use of expressions involving a small number of symbols to represent all the concepts in the problem can speed up the solution to the problem by immediately reducing the number of unknowns to those having only nontrivial relations to one another. However, it must be recognized that you are performing two steps at once: representing the concepts in the problem and expressing some relatively simple relations between concepts. Trying to do these two steps at once increases the probability of error, though when performed successfully, it usually allows you to solve the problem faster. My advice is to try to combine these two steps and use expressions to represent concepts. However, if you find yourself making lots of errors, go back to the more compulsive and systematic procedure of first assigning symbols to all different and important concepts in the problem, and only then start expressing the relations between these concepts — for example, by equations such as $j = b + 30$, where j represents John's weight.

Subscripts

In a problem with many different quantities of the same type — such as many different times, rates, distances, volumes, heights, or radii of circular bases — it is often desirable to use two-symbol codes for representing these concepts. In these problems, the *type* of concept can be thought of as a *property* of some object, activity, or object-performing-an-activity. One of the two symbols in the two-symbol code is used to represent the property and the other symbol to represent the object, activity, or object-performing-an-activity. In addition, the mnemonic convention is that the property is represented by the main symbol, with the object, activity, or object-performing-an-activity represented by a subscript. So, for example, the height of cylinder A could be represented by h_A and the radius of its base by r_A. Similarly, h_B and r_B could be used for cylinder B, and so on.

Subscripts usually appear to the right and somewhat below the main symbol, but they occasionally appear to the lower left of the main symbol. So long as there are no exponents involved in solving a problem, you can use superscripts for the second symbol just as well as subscripts. Superscripts can appear to the upper left or upper right. However, since there is very frequently a danger that superscripts will be confused with exponents, it is good practice to avoid using them.

The only possible excuse for using a superscript occurs in problems where the objects to which a property applies are differentiated along two or more dimensions, and the values of the various dimension are completely noncomparable. For example, imagine that in some problem you have various complex containers, each of which has several component containers with different shapes and dimensions. Container A might be composed of a cylindrical subcontainer, two cubical subcontainers, and a rectangular subcontainer. There are also complex containers B and C, each composed of subcontainers. How should you represent the volume of each subcontainer in container A, for example? For ease in remembering what symbols represent what and the associated ease in retrieving formulas for volumes from your long-term memory, you might try using a notation such as the following: Let V_c^A be the volume of the cylindrical subcontainer of A; let $_1V_s^A$ be the volume of the first cubical (square base) subcontainer of A; let $_2V_s^A$ be the volume of the second cubical subcontainer of A; let V_r^A be the volume of the rectangular subcontainer of A.

You encounter notations like the above, with both left and right subscripts and a superscript (or even two!), but I think they are poor notations. Superscripts are dangerous because they can be confused with exponents. If a superscript is put to the upper left of the main symbol, it eliminates the possibility of it being confused with an exponent for the main symbol; but it is then apt to be mistaken for an exponent for the symbol to the left in a multiplication problem (for example, h^AV_r for "h times ^AV_r"). If you are always careful to put the entire three- or four-symbol complex into parentheses—for example, $(_2V_s^A)$—you will avoid these confusions. However, you will then have a complex symbol with as many as six component symbols; indeed, even four component symbols are too many.

In cases like the above, you should probably scrap the entire effort and use a single letter for each concept or, at most, a two-symbol code for each concept. If you can easily think of some single symbols that have some reasonably good mnemonic relation to the concepts they represent, use such symbols. Or, failing this, make an arbitrary assignment of single symbols to concepts and write down the assignment in a table or diagram, for consulting when necessary. The time saved in the repeated writing of the symbols in various statements or equations usually more than compensates for the extra time spent in table or diagram lookup. Furthermore, it is rather hard to visualize, verbalize, or otherwise think about such complex symbols as the ones in the above example.

Multiple Subscripts

One important exception to the above advice regarding complex symbols are cases where you have an entire matrix (two-dimensional or higher dimensional) of objects or activities, each entry of which has one or more properties. In such cases, use multiple subscripts (possibly separated by commas), all on the lower right of the main symbol (for example, x_{ijk} or $x_{i,j,k}$). Such cases arise frequently in statistics, where

x_{ijk} might represent the wheat yield on the kth plot of land, subjected to the ith value on one treatment dimension (for example, the amount of some kind of fertilizer), and subjected to the jth value on another treatment dimension (for example, the amount of water).

Why is use of a complex symbol with multiple subscripts recommended in the statistics example and not in the previous example? In the statistics example there are a number of dimensions, and every combination of values on every dimension (that is, every entry in the matrix) has a defined value of the property in question (for instance, wheat yield). In the problem concerning volumes of complex containers, every complex container could have different shapes of subcontainers and different numbers of each shape. Suppose you defined dimensions like the "shape of a subcontainer"; the complex container to which the subcontainer belonged; and whether it was the first, second, and so on, subcontainer of this shape to be a part of the complex container. You would then have lots of cells in the matrix with no objects corresponding to them in the problem (empty cells in the matrix). This is more trouble than it is worth.

When you actually have a large number of objects in some multidimensional matrix, there is really no feasible alternative to using multiple subscripts. Furthermore, in matrix problems like the statistics example, we often do operations such as summing over all the entries in some row or column without ever having to look up in any table or diagram the meaning of each complex symbol. When subscript notation is used, a convenient notation exists for indicating sums or products—for instance:

$$\sum_{i=1}^{n} x_{ijk} \quad \text{and} \quad \prod_{i=1}^{n} x_{ijk}$$

Subscript notation is always indicated in problems where multiple sums or products of this type are likely to be used in the solution. Where such multiple sums and products or some other computations or relations involving the subscripts are not likely to appear anywhere in the solution, it is questionable whether you should ever use a symbol with more than one subscript.

Example Problem

Tom, Dick, and Harry mow lawns in the summer to earn money. They each have a lawn mower, and one Saturday they decide to mow a 5,900 square foot lawn together, using all three lawn mowers. Tom mows 70 square feet per minute, Dick 50, and Harry 40. Dick and Harry start mowing the lawn at the same time, but Tom has trouble starting his mower and is delayed for 30 minutes. All three boys stop mowing at the same time, when the lawn is finished. How long does Tom mow?

In solving this problem, the principal step is to represent the information in algebraic notation and set up the equations. After this, the sequence of algebraic actions (operations) is trivial. Stop reading, and represent the information in this problem, using the principles discussed in this chapter; then solve the problem.

There are several steps involved in representing the information, and you should be aware of them, even though you may be able to solve such problems quickly and easily. The same types of steps are involved in all story-algebra problems, and similar representational steps are involved in many other types of problems as well.

First we have to represent the unknown quantities for which we are to solve. Here that means having some expressions represent (stand for) the times that Tom, Dick, and Harry work. It is economical and an aid to visual and verbal memory to choose a single symbol (often a letter) to stand for an unknown quantity. You may choose any symbols you like for the unknown quantities represented, provided that you do not use the same symbol to represent quantities that might not be equal.

It is perfectly safe to use different symbols for quantities that may later prove to be equal. However, it aids the memory, if you notice such equalities before you assign symbols to quantities and assign the same symbol to quantities that must be equal. For example, in the present problem, we can use t for the time that Dick mowed and also for the time that Harry mowed. We know that these times are equal since Dick and Harry start and stop at the same time. It would place an unnecessary strain on the memory to use t_0, t_1, and t_2 to stand for the mowing times of Tom, Dick, and Harry, respectively, though there is no mathematical reason why numerical subscripts cannot be used for these purely nominal or naming purposes.

Note that there is a good mathematical reason for not choosing to represent Tom's mowing time by the symbol 8, because all number symbols, including 8, are already implicitly given as concepts in a story-algebra problem, and Tom's mowing time is not known to be equal to 8. This sort of objection does not apply to the use of numbers in a purely nominal way in subscripts, and, of course, nominal numerical subscripts are used frequently in mathematics, science, and engineering. Often the problem gives only the names "first force" or "second force," and in such cases the obvious representation is f_1 and f_2.

In problems with many unknown quantities, it is desirable, for the same reason of minimizing the load on human memory, to choose letters that have easy mnemonic relations to the full names of the quantities represented. For example, we choose t or T for time quantities and r or R for rate quantities. If there are many different quantities of the same type, such as several different times, then it may be best to represent them by a two-symbol code, such as t_T, t_D, and t_H, where the t indicates a time and the subscript indicates which time. The subscript should also have some easy mnemonic relation to the full name of the quantity represented. In the present problem, we might initially have chosen t_T, t_D, and t_H to stand for the times that Tom, Dick, and Harry mow, though, as already noted, this is unnecessary.

Now here is the solution. This is a simple work-rate problem. The primary equation to use in such problems is that work equals the sum of all the rate times time components. In this problem, that means setting up an equation such as

$$70(t - 30) + 50t + 40t = 5,900$$

where
$$t = \text{time Dick and Harry mow}$$
$$t\text{-}30 = \text{time Tom mows}$$

Solving the above equation gives $t = 50$, and therefore Tom mows for 20 minutes.

SOME IMPORTANT MATHEMATICAL CONCEPTS

Ordered and Unordered Pairs

Ordered pairs without replacement   Each of the possible *permutations* of m things taken two at a time is just another name for an ordered pair of elements (i,j), such that both i and j are members of the set of m things and i and j are *not* the same thing or element of the set. This is sampling twice from a set of m elements *without* replacement of the element already sampled. Ordered pairs of this type (permutations) are frequently involved in problems.

For example, consider a problem in which some group has the ritual of everyone kissing everyone else on the forehead, and you are supposed to determine the number of people in the group from the number of kisses or vice versa. Kissing on the forehead is represented by an ordered pair of persons in which the first member of the pair is the kisser and the second member is the person kissed. Having recognized that this is an ordered pair (permutation) problem, you can solve the problem in a manner similar to the way you solved other problems— namely, by determining how many ways there are to fill each position in the ordered pair. In this problem, there are m ways to fill the first position, and, with it filled, there are $m - 1$ ways to fill the second position. Hence, there are $m(m-1)$ ways to fill both positions (with different elements). Thus, the m persons exchange $m(m-1)$ forehead kisses. Of course, to determine m from $m(m-1)$, you have to solve a quadratic equation, but in the average problem of this type, that would be very simple.

Ordered pairs with replacement A known or unknown point in a plane is often best represented by an ordered pair of symbols representing its distances from each of two (usually perpendicular) axes, though, of course, in many problems a point can be represented by just a single symbol such as A or a dot on a page. However, frequently the representation of a point in a plane should be by an ordered pair of symbols — an ordered pair of numbers for a known point, an ordered pair of letters for an unknown point. Since the values of the two coordinates can be equal, the representation of a point in a plane by an ordered pair is an example of sampling twice *with replacement* from the entire population of coordinate values (whether finite or infinite). Sampling with replacement means that when you have chosen the value for the first coordinate, you put that value back into the population so that it can be drawn again as the value for the second coordinate. Thus, if there were m possible coordinate values, there would be m^2 possible points — that is, m^2 ordered pairs of coordinate values.

Unordered pairs without replacement Although forehead kisses are represented by ordered pairs of persons, lip kisses (assuming mutual kissing) are represented by unordered pairs. That is, if A kisses B, B is assumed also to be kissing A. Thus, there is no basis for distinguishing between (A, B) and (B, A).

This is a combinations problem rather than a permutations problem. As we discussed in Chapter 9, the number of combinations of m things taken two at a time is $m(m-1)/2$. We derived this figure by reasoning as follows: The number of permutations is $m(m-1)$, and there are two permutations for every combination (two ordered pairs for each unordered pair). Thus, we should divide the number of ordered pairs (permutations) by 2 to get the number of unordered pairs (combinations).

A line segment is determined by its two end points. Often a line (segment) will be defined as an unordered pair of (distinct) points, with every different unordered pair of points in the total set of points representing a different line. Notice that a line is indeed an unordered pair of points, not an ordered pair, in most cases. However, it is perfectly possible to have *directed* line segments in a problem (that is, lines with arrows on one end, or *vectors*), in which case the line (A, B) is different from the line (B, A), where A and B are two different points in the set of points.

The above two examples of unordered pairs are instances of unordered pairs where the sampling is without replacement. A person is not assumed to be able to kiss himself on the lips. A line segment is determined by two end points that are distinct, that is, two different points.

Unordered pairs with replacement We can also obtain unordered pairs by sampling from a population with replacement. An example might be the distinct combinations obtained by throwing two dice, which might produce these results: 6-6, 6-5, 6-4, 6-3, 6-2, 6-1, 5-5, 5-4, . . . , 2-2, 2-1, 1-1. To compute the number of distinguishable throws of two dice, we reason that the order of the two dice is not important and thus 6-5 is the same outcome as 5-6.

Offhand you might think that the number of unordered pairs obtained by sampling with replacement would be equal to the number of ordered pairs sampling with replacement divided by 2, as was the case for sampling without replacement (permutations and combinations). However, this is not the case. Most of the unordered pairs of outcomes obtained by sampling with replacement do indeed have two distinct ordered-pair counterparts, but there is a subset of the unordered pairs each of which has only a single ordered-pair counterpart. These latter pairs are 6-6, 5-5, 4-4, 3-3, 2-2, and 1-1.

In general, if you are sampling with replacement two times from a population of m elements, the number of different unordered pairs will be

$$m + \frac{m^2 - m}{2} = \frac{m(m + 1)}{2}$$

This result is obtained by reasoning as follows: exactly m of the unordered pairs are of the form i-i, where $i = 1, \ldots, m$. The remaining $m^2 - m$ ordered pairs have exactly two ordered-pair counterparts for each unordered pair. Thus, the quantity $(m^2 - m)$ should be divided by 2 to get the number of unordered pairs, and m should be added to this total to get the total number of different unordered pairs.

Systematic listing of unordered pairs Occasionally we need to list all the unordered pairs that can be obtained by sampling, either with or without replacement, from some population. The efficient way to accomplish such a listing is to put all of the elements in the population into an ordering, whether they have any natural ordering or not. Having ordered the elements from 1 to m, we can then list all of the unordered pairs by proceeding as follows: If the unordered pairs are being obtained by sampling with replacement, we take each element from the population and pair it with itself and every element below it in the ordering. Thus, we would obtain pairs such as 5-5, 5-4, 5-3, 5-2, 5-1, 4-4, 4-3, 4-2, 4-1, 3-3, 3-2, 3-1, 2-2, 2-1, 1-1. If the unordered pairs are being obtained by sampling without replacement, we take each element in the population and pair it with each of the elements below it in the ordering. Thus, we would have listings such as 5-4, 5-3, 5-2, 5-1, 4-3, 4-2, 4-1, 3-2, 3-1, 2-1.

Importance of ordered and unordered pairs   Ordered and unordered pairs are very common concepts in problems, and you should be alert for the possibility of representing concepts as ordered or unordered pairs of other concepts, frequently an important step in solving the problem. In a sense, this representation involves your recognizing a relation between different concepts and incorporating that relation into your representation of concepts in the problem, which is similar to representing concepts by expressions, as we discussed previously. You should also be careful to note whether the ordered and unordered pairs are being obtained with or without replacement, for, as stated earlier, it makes a considerable difference in the number of such pairs.

Often you will realize that some concept can be represented by a pair of other concepts before you realize whether the pair should be considered an ordered pair or an unordered pair and whether the sampling is with or without replacement. However, by being explicitly aware of the distinctions, you will try to decide whether the pair is ordered or unordered and whether the sampling is with or without replacement, before trying to use the concept in solving the problem.

Ordered and Unordered Sets

Ordered and unordered pairs generalize easily to ordered and unordered sets of elements greater than two. So we can have an ordered or unordered set of three elements (A, B, C), four elements (w, x, y, z), and so on.

Ordered sets without replacement   Each permutation of m things taken n at a time $(n < m)$ is an ordered set of n elements. In getting any particular permutation, we are sampling n times without replacement from a set of m elements. Thus, there are m possible ways to fill the first position (m possible elements that could be selected first), $(m - 1)$

ways to fill the second position, . . . , and $(m - n + 1)$ ways to fill the nth (last) position, or $m(m - 1) \ldots (m - n + 1) = m!/(m - n)!$ different ways to select all n elements (assuming all m elements are distinct). An example of a permutation problem (ordered sets, sampling without replacement) is as follows:

> A gym teacher wishes to put on a balancing demonstration in which one of the stunts will be to have four boys stand on each others shoulders in a single tower. Out of the class of 20 boys, the gym teacher wishes to select the most stable tower of four boys. To do this he plans to try each possible tower of four boys once and time how long they are able to balance successfully on each others shoulders without falling over. How many such towers of four boys must the gym teacher investigate?

Stop reading and try to solve the problem.

Since there are 20 boys in the gym class and a tower of four boys constitutes an ordered set of four elements sampled from the class without replacement, this is a permutations problem. Therefore, the number of possible towers is $20!/(20 - 4)! = 20!/16!$.

Ordered sets with replacement A point or vector in n-dimensional space can be represented by an ordered set of its n coordinates $(x_1, x_2, \ldots, x_n)$. Such an ordered set of n elements is obtained by sampling with replacement from a population of, say, m possible coordinate values exactly n times. The number of possible ordered sets obtained by sampling n times from a population of m elements is equal to m^n.

Unordered sets without replacement A triangle is an unordered set of three different points (sampling three times without replacement from the set of all points). A quadrilateral is an unordered set of four different points in a plane. Each combination of m things taken n at a time $(n < m)$ is an unordered set of elements such that none of the elements is identically the same element (for example, one is sampling n times without replacement from a set of m elements). The number of combinations of m things taken n at a time is simply the number of permutations of m things taken n at a time, divided by the number of different permutations for the same combination of n elements. Since there are $n!$ different permutations for each combination of n elements, there are $m!/[n!(m - n)!]$ combinations of m things taken n at a time. An example of a combinations problem is as follows:

> How many different bridge hands (13 cards) can be obtained by dealing 13 cards out of a standard 52-card deck?

Dealing 13 cards from the standard 52-card deck is sampling without replacement. The order in which the cards are dealt to someone makes no difference in defining a bridge hand. Therefore, this is a combinations problem, that is, a problem involving an unordered set obtained by sampling without replacement. Thus, the number of bridge hands is $52!/13!(52 - 13)! = 52!/13! \ 39!$.

Unordered sets with replacement   Thus far we have discussed ordered sets obtained by sampling with and without replacement and unordered sets obtained by sampling without replacement. Computing the number of unordered sets obtained by sampling with replacement is a far less trivial problem. The problem has an extremely elegant solution, which I found in Feller (1957, p. 38) and which I think provides a good example of how clever representation of the information in the problem can facilitate its solution. Thus, we will examine the solution from two viewpoints—that of determining the formula for the number of unordered sets obtained by sampling with replacement, and that of having an elegant example of how to represent information in a particular class of problems.

The basic problem is to determine how many unordered sets we can obtain by sampling n times with replacement from a population of m elements. The solution is obtained easily by considering the m elements of the population, to be represented by the *spaces* between $m + 1$ boundary markers ordered along a line. That is, we will imagine we have a line with $m + 1$ interval boundaries marked off along that line defining m intervals, as shown in Fig. 10-2. In the figure, the n elements sampled are represented by *circles* between various boundary markers. The number of circles between the first and second boundary marker represents the number of times the first element was sampled. The number of circles between the second and third boundary marker represents the number of times the second element in the population was sampled, and so on. With this representation, we can easily compute the number of different unordered sets that can be obtained by sampling n times without replacement from a population of m elements. The two end boundary markers out of the $m + 1$ boundary markers must remain fixed at the ends. The remaining $m - 1$ boundary markers and n elements sampled can be rearranged at will. If we con-

| O O | O O O | | O O O O | | | O |

FIGURE 10-2
Clever reformulation of the information in the problem of determining the number of unordered sets that can be obtained by sampling n times from a population of m elements.

sider any particular rearrangement to be obtained by sampling *n* times *without* replacement from the population of $n + m - 1$ elements, then the number of such rearrangements of the lines and circles is easily determined, namely, $(n + m - 1)!/n! \, (m - 1)!$.

Thus, we have transformed the problem of sampling *n* times with replacement from a population of *m* elements to the problem of sampling *n* times without replacement from a population of $n + m - 1$ elements. In both cases, we are trying to compute the number of unordered sets obtained by such a sampling. Since we know the solution to the problem of obtaining the number of unordered sets obtained by sampling without replacement (combinations), we now know the answer to the problem of determining the number of unordered sets obtained by sampling with replacement.

Relations

Relations are labeled connections between concepts. Examples of relational concepts include, among others: father of, brother of, sibling of, descendant of, prior to, less than, equal to, identical to, heavier than, older than, beside, includes, is a member of. Relations can be written as "*a* R *b*" (meaning perhaps "*a* is the father of *b*") or as R(a, b). In the latter case, R(a, b) is true (equals 1), if and only if the relation R obtains for the ordered pair (a, b), and R (a, b) is false (equals zero), if and only if the relation R does not obtain for the ordered pair (a, b).

Relations can be classified according to whether they satisfy certain properties (axioms). For example, a relation can be reflexive (*a* R *a*, for all *a* in some set), antireflexive (*a* not-R *a*, for all *a* in some set), or neither. Relations can be symmetric or commutative (*a* R *b* implies *b* R *a*, for all *a* and *b* in some set), antisymmetric (*a* R *b* implies *b* not-R *a*, for all *a* and *b* in some set), or neither. Relations can be transitive (*a* R *b* and *b* R *c* imply *a* R *c*, for all *a*, *b*, and *c* in some set) or not. Relations that are reflexive, symmetric, and transitive form an especially important class of relations known as *equivalence relations*. "Equal to " and "identical to" are equivalence relations, but so are "has the same weight as," "is the same color as," and "is just as good as."

Some students find it difficult to distinguish the concepts of *equivalence*, *equality*, and *identity*, and indeed the meanings of these concepts are somewhat variable, especially in going from science to mathematics. To get an idea of why we sometimes need to distinguish them, consider the following examples.

Suppose you were thinking about your prospects for immortality, and you imagined that it might be possible to form a complete duplicate of yourself that was the same configuration of molecules as yourself but, of course, used different molecules. Such a duplicate would be equal to you in every respect, but the duplicate would not be identical to you, since the two of you in fact would be two different parts of the universe. Identical twins are considered to be genetically equal, but they are not identical, since they are two different entities. All the molecules of some class are considered to be equal, but since there is more than one such molecule in each class, the molecules of a given class are not identical. A thing is identical only to itself, but it can be equal to all duplicates. Of course, many people feel that it is likely that no two entities can be *exact* duplicates in every respect, but this is not too important for the definition of equality. We can simply say that two things are equal, if there is no property that we can currently determine to distinguish them (except of course the fact that they are not the identically same entity).

Two entities may be *equivalent* to each other in some one respect (one property) without being *equal* to each other (equivalent to each other in all respects). This is pretty obvious when it is put this way. Two girls can have the same weight (to the nearest pound), have the same shooting percentage in basketball, have the same number of points on a test, and so on. Thus, it is clear that (a) two names of objects that are identically the same really refer to the same object, (b) two names of objects that are equal refer to two different objects that are equivalent in all respects, and (c) two names of objects that are equivalent in some respect are not necessarily identical or equal in all respects.

Sometimes there is no need to distinguish all three types of "sameness" concepts. For example, in real-variable mathematics, where a symbol has only one property, its numerical value, there is no need to distinguish equivalence and equality. Furthermore, our definition of identity is not quite the same intuitively as it is when we discuss objects presumed to exist in the real world. In mathematics not viewed as applying to the real world, two expressions are identically equal ($\equiv$), if and only if they have the same values across *all* substitutions for free variables (for example, $x^2 - 1 \equiv [x + 1][x - 1]$). Two expressions are often said to be equal ($=$) if and only if they have the same values for at least one substitution for free variables (for example, $x^2 + 2 = 6$ for the substitution $x = 2$).

Operations

In Chapter 2, operations were contrasted with givens. Intuitively, operations were the things that you could do to change the state of the problem and givens were the materials you had to work with (the starting point or initial problem state). You could attach a symbol to each operation that you could perform to change the problem state, but what would that accomplish? For one thing, it might allow you more conveniently to list the alternative actions that could be performed at each node in the state-action tree. However, this is not the primary reason for attaching symbols to operations.

The primary reason for using a symbol to represent an operation is that you can formulate problems in which the givens are composed of statements that involve action concepts as well as object or property concepts. Consider a problem to solve two linear equations in the variables x and y for the values of x and y—for example, $2x + y = 1$ and $x - 6y = 20$. Addition, subtraction, and multiplication operations are indicated in each of these statements, but the particular actions they indicate (for example, multiply y by 6) are not the actions you take in solving the problem. The actions you take are to add, subtract, multiply, or divide both sides of some equation by the same quantity and to substitute equals for equals.

By contrast, there are problems in which you might take the operations of addition, subtraction, multiplication, and so on, of two quantities as the operations used at various nodes of the state-action tree. For example, water-jar problems (such as the one discussed in Chapter 8) involve presenting several jars that hold different quantities of liquid. One is asked to produce some quantity of liquid that is different from the capacity of any jar. The method of solution involves, in essence, adding and subtracting the capacities of each of the jars.

Imagine that you have a five-quart jar and a three-quart jar and are attempting to obtain exactly one quart of water. You could fill the three-quart jar, pour it into the five-quart jar, fill the three-quart jar, pour two quarts into the five-quart jar, and have exactly one quart left in the three-quart jar, as required.

In water-jar problems, adding and subtracting quantities are the operations in the state-action tree. In solving equations, adding and subtracting quantities are operations used in constructing statements; adding and subtracting the same quantities *to both sides of an equation* are operations in the state-action tree. Perhaps there should be two different names to distinguish operations at these two different levels in a problem. In any event, it is important for clear thinking to keep operations at the two levels distinct in your mind.

Mappings and Functions

Imagine that you have the elements of some set A (the argument set), the elements of another set T (the target set), and a set of arrows, each going from one member of the argument set to one member of the target set. Not all of the elements of set A need have arrows going from them. Some elements of set A may have several arrows going from them to different elements in set T. Some of the elements of set T may have no arrows going into them, and some may have several arrows going into them. Any such system of arrows linking two sets is called a *mapping*. Although it is easier to explain mappings in diagrammatic (spatial) terms, a mapping can be represented verbally as a *set* of ordered pairs where the first member of the pair is an element from set A and the second is a corresponding element in set T. The number of ordered pairs in the set is the same as the number of arrows in the diagrammatic representation. Of course, most mappings of interest are represented more simply by a rule that allows us to compute the element in T associated with each element in A.

One example of a mapping is illustrated in Fig. 10-3, which suggests that sets A and T are completely distinct, that is, have no elements in common (are nonoverlapping). This need not be the case. Sets A and T could be identical sets, either one could be completely included in the other, they could be overlapping (some elements common to both sets, but some elements in each set being not contained in the other set), or they could be nonoverlapping. In short, any set relation is

Set A Set T

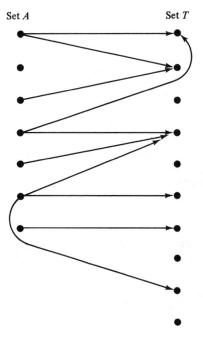

FIGURE 10-3
A mapping from set A to set T, where set A and set T are nonoverlapping.

possible for sets A and T. An example of a mapping involving two overlapping sets is shown in Fig. 10-4.

In addition to the relation between the argument set (set A) and the target set (set T) as just discussed, mappings have other properties. A mapping is *complete* if and only if it is defined over all members of the argument set (that is, all members of set A have an arrow going from them); otherwise it is *incomplete*. Often we assume that all mappings are complete, since the argument set could always be reduced to those elements for which the mapping is defined, with no loss of information regarding the mapping other than that regarding its incompleteness.

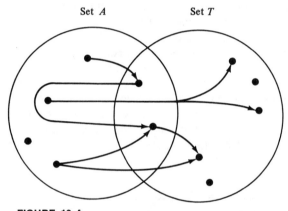

Set A Set T

FIGURE 10-4
A mapping from set A to set T, where set A and set T are overlapping. In this case, the sets have two common elements. Sets are enclosed in circles. Common elements are in the overlapping part of the two circles.

A mapping is *single valued* (is *well defined* or is a *function*) if and only if each element in the argument set maps into no more than one element in the target set (that is, there is no more than one arrow going out of each element in set A). If a mapping is both single valued and complete, then exactly one arrow goes out of each element in set A. The term *function* is sometimes reserved only for single-valued mappings, but frequently the expression *multivalued function* is also heard, which indicates that there is not complete consistency in the restriction of the term function to single-valued mappings. An example of a complete and single-valued mapping is shown in Fig. 10-5.

A mapping is an *onto* mapping if and only if every element of the target set has an arrow going into it. Another way to say this is that the

Set *A*   Set *T*

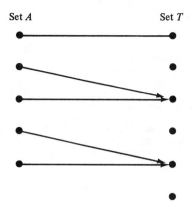

FIGURE 10-5
A single-valued mapping from set *A* to set *T*. The mapping is also complete since all members of set *A* have arrows going from them. Note that the mapping is single-valued in going from set *A* to set *T* but not in the inverse direction, nor is the inverse mapping even complete (arrows do not go into every element in set *T*).

inverse mapping (going backward along the arrows from set *T* to set *A*) is complete (that is, defined over every element in set *T*). An example of an onto mapping is shown in Fig. 10-6.

A mapping is *one-to-one* if and only if every element of the target set has no more than one element going into it. Another way to say this is that the inverse mapping (going backward along the arrows from set *T* to set *A*) is single valued. Note that a one-to-one mapping need not be onto, just as a single-valued mapping need not be complete. That is, the one-to-one property requires that no more than one arrow go into each element in the target set, whereas the onto property requires that at least one arrow go into each element in the target set. Similarly, the single-valued property requires that no more than one arrow go out from each element in the argument set, whereas the completeness property requires that at least one arrow go out from each element in the argument set. An example of a one-to-one mapping that is not single-valued, onto, or complete is shown in Fig. 10-7.

Set *A*   Set *T*

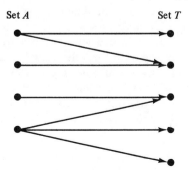

FIGURE 10-6
An onto mapping from set *A* to set *T*, meaning that the inverse mapping from set *T* to set *A* is complete (defined over all elements in set *T*). The mapping from set *A* to set *T* is also complete but is not single valued.

Set *A*   Set *T*

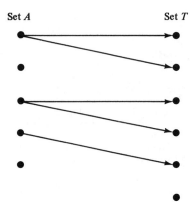

FIGURE 10-7
A one-to-one mapping from set *A* to set
T that is not complete (two elements of
set *T* have no arrows going from them),
not onto (one element of set *T* has no
arrow going into it), and not single
valued (some elements of set *A* have
more than one arrow going from them
to different members of set *T*).

Real-Valued Functions of a Real Variable

The most commonly encountered functions are real-valued functions
of one or more real variables (real arguments). A statement such as
$y = f(x)$ means that an element *x* is mapped into an element *y* according
to the rule (function) represented by the symbol *f*. In this section, we
are assuming that *x* and *y* are real numbers (the ordinary positive and
negative integers, fractions, square roots of positive integers, and so on,
with which we are familiar). Examples of functions that *f* could represent include polynomials with known coefficients ($y = 7x + 2$, $y = 4x^3$
$- 3x^2 + x - 37$), polynomials with unknown coefficients ($y = ax^4 - bx$
$+ c$), and trigonometric functions ($y = \sin x + a \tan^3 x$).

In some problems, you may know that one variable is a function of
one or more other variables, but it may take some problem solving
to determine exactly what the function is. In such cases, it is generally
helpful to assign some symbol like *f* to the unknown function and
write equations involving the unknown function [for example, $y = f(x)$].

In some problems, you can reduce the number of symbols you have
to remember by just writing an expression such as $y = y(x)$, meaning
that the variable *y* is a function of the value of *x*, but without bothering
to give a separate name to the function (separate from the value of the
function when *x* is the argument). This is a useful mnemonic trick in
simplifying notation in problems where there is no possibility of confusing the concept of the function (*f*) with the concept of the dependent
variable (*y*). However, when such confusion is possible, this trick
should be avoided.

11

Problems from Mathematics, Science, and Engineering

This chapter is designed to establish the generality of the problem-solving methods discussed throughout the book. In previous chapters, the problems used to illustrate the methods were deliberately selected so that they could be solved by the reader with no more background than a high school student with one year of algebra and one year of plane geometry. Many of the problems were of the puzzle (or brain teaser or recreational mathematics) variety, which require no specialized knowledge of mathematics, science, or engineering. Although methods for solving such problems have some recreational interest, there is also a serious practical reason in mastering them, for they are also useful for solving serious problems in all areas of mathematics, science, and engineering. This chapter is designed to demonstrate this applicability and to give the reader some experience in it.

ALGEBRA

The solution of systems of simultaneous linear equations provides a simple example of the use of evaluation functions, hill climbing, and subgoals. As an example, consider the following system of three linear equations:

$$2x + y - 3z = 1 \qquad \text{(E1)}$$

$$x + 2y + 5z = 9 \qquad \text{(E2)}$$

$$3x - 3y - 10z = 4 \qquad \text{(E3)}$$

The operations available for solving such a system are essentially the following. We can (a) multiply both sides of an equation by the same number, (b) add equals to equals (or subtract equals from equals), and (c) substitute equals for equals. As an example of the first, consider the action of multiplying both sides of equation (E2) by the number -2. This yields the equation $-2x - 4y - 10z = -18$. As an example of the second operation, consider the action of adding the equation just derived from (E2) to (E1):

$$2x + y - 3z = 1 \qquad \text{(E1)}$$

$$\underline{-2x - 4y - 10z = -18} \qquad (-2) \cdot \text{(E2)}$$

$$-3y - 13z = -17$$

or

$$3y + 13z = 17 \qquad \text{(E4)}$$

As an example of the third operation, consider the following substitution of an expression for x, derived from (E2), into (E1):

$$x = -2y - 5z + 9 \qquad \text{(E2)}$$

$$2(-2y - 5z + 9) + y - 3z = 1 \qquad \text{(E1)}$$

If we were given a particular numerical value for x, we could, of course, also substitute that particular value for x anywhere it appeared in any equation. The goal is to derive three expressions of the form $x = \underline{\hspace{1cm}}$, $y = \underline{\hspace{1cm}}$, and $z = \underline{\hspace{1cm}}$, where specific numbers appear in the blanks.

Now stop reading and try to solve the problem.

The solution of such a problem involves primarily the use of an evaluation function and the subgoal method, with perhaps a little use of hill climbing. The evaluation function is concerned with the number of variables (unknowns) in each equation and the number of independent equations involving any particular set of variables (unknowns). The original system of equations consists of three equations, each of which has three variables (unknowns). From this starting point, a more highly valued state would be one in which we had two equations involving the same two unknowns. An even more highly valued state would be one in which we had one equation involving one unknown. Still more highly valued would be a state in which we had two equations, each of which involved a different, single unknown. The most highly valued state of all—short of solution—would be one in which we had three equations, each involving a different, single unknown. For the purpose of defining the present state-evaluation function, note that we have ignored the subproblem of solving a single linear equation with one variable for the value of the unknown, since we assume that to be a trivial subproblem whose method of solution is already well known. We have not bothered to assign numbers to states that have the above-mentioned properties, because this is unnecessary for solving this problem. There are several ways to assign specific numbers to these states, and any of them would be equally satisfactory as a guide to the definition of successive subgoals in solving the problem.

In learning to solve systems of linear equations by means of the above three operations, you should first master the solution of linear equations involving one unknown, then systems with two independent equations involving two unknowns, then three independent equations involving three unknowns, and so on. You should learn that the first subgoal to achieve in a system of n equations with n unknowns is to derive a system of $n - 1$ equations involving $n - 1$ unknowns. The next subgoal is to derive a system of $n - 2$ equations involving $n - 2$ unknowns, and so forth. Occasionally, it is possible to jump several levels at once, and this is even better, but in general you must proceed one step at a time. Now stop reading and solve the problem, if you did not do so before.

To solve the above problem, we should first set a subgoal that we must achieve a system involving two equations and two unknowns. Somehow, then, we must derive two equations, from each of which we have eliminated the same unknown. Since two such equations must be derived, there are two parts to this first subgoal (two subgoals of the first subgoal). There are a variety of ways to accomplish the first subgoal, one of which is as follows:

$$2x + y - 3z = 1 \qquad \text{(E1)}$$
$$\underline{-2x - 4y - 10z = -18} \qquad (-2) \cdot \text{(E2)}$$
$$ - 3y - 13z = -17$$

or

$$3y + 13z = 17 \qquad \text{(E4)}$$
$$3x - 3y - 10z = 4 \qquad \text{(E3)}$$
$$\underline{-3x - 6y - 15z = -27} \qquad (-3) \cdot \text{(E2)}$$
$$ - 9y - 25z = -23 \qquad \text{(E5)}$$

Having achieved the first subgoal, the next subgoal is to solve this system of two equations and two unknowns to derive a single equation involving one unknown, as follows:

$$-9y - 25z = -23 \qquad \text{(E5)}$$
$$\underline{9y + 39z = 51} \qquad (3) \cdot \text{(E4)}$$
$$14z = 28$$
$$z = 2$$

Having achieved the second subgoal (including finding the value of one of the unknowns), it is time to proceed to the third subgoal of deriving another single equation involving a single unknown. This derivation can be done by using the substitution operation, as follows:

$$3y + 13 \cdot 2 = 17 \qquad \text{(E4)}$$
$$3y + 26 = 17$$
$$3y = -9$$
$$y = -3 \qquad \qquad *$$

Finally, the fifth subgoal and the final component in the solution of the problem is as follows:

$$x + [(2) \cdot (-3)] + [(5) \cdot (2)] = 9 \qquad \text{(E2)}$$
$$x - 6 + 10 = 9$$
$$x = 5 \qquad \qquad *$$

To solve the above problem, you have to know where you want to go at all stages in its solution. That is, you must have an evaluation function similar to that discussed here. The evaluation function provides the means for defining a series of subgoals (subproblems) that lead to the solution of the entire problem. Along the way, in the achievement of some of the subgoals, one equation might be multiplied by a number to yield an equation with the same coefficient for a particular unknown as some other equation already obtained. This action illustrates the use of another evaluation function—namely, getting two equations to involve the same coefficient for a particular variable. Since achieving this subgoal is relatively simple, we might view the selection of the appropriate action to achieve this subgoal as hill climbing. However, I think that viewing the solution in terms of the subgoal method is far more accurate and important.

Now let us consider the solution of a very different type of equation:

$$4^{x-3} = 2^x \cdot 3^{x+1}$$

The goal is to derive an equation of the form $x = \underline{\hspace{1cm}}$. The operations available include all those specified in the previous problem. Also available are operations that may be stated generally as "doing the same thing to both sides of an equation": adding the same number to both sides, subtracting the same number from both sides, multiplying or dividing both sides of the equation by the same number, raising both sides to the same power, taking the same root of both sides, or taking logs of both sides. (Remember, however, that operations that increase the degree of an equation will add roots and operations that reduce the degree of an equation will subtract roots.) For the purposes of solving this problem, the only properties of logarithms that we need to know are that $\log (a^b) = b \log a$ and that $\log (a \cdot b) = \log a + \log b$.

Now stop reading and try to solve the problem.

Since one property of the goal expression is that the x does not appear in an exponent, one subgoal that can be defined immediately is to derive an equation in which x does not appear in an exponent. Stop reading and try to solve the problem, if you did not before.

This sort of problem would appear in conjunction with a chapter on logarithms, since to achieve the subgoal we must take logarithms of both sides of the equation, as follows:

$$(x - 3)\log 4 = x \log 2 + (x + 1)\log 3$$

$$(\log 4 - \log 2 - \log 3)x = \log 3 + 3 \log 4$$

$$x = \frac{\log 3 + 3 \log 4}{\log 4 - \log 2 - \log 3}$$

Of course, we must know the relevant properties of logarithms in order to solve this problem. In addition, we must define the subgoal of achieving an equation that is in a form for which we know the appropriate solution methods, just as in the case of simultaneous linear equations. In virtually all problems from mathematics, science, and engineering, there is an interplay between the use of specialized knowledge and the use of general problem-solving methods. Either the lack of specialized knowledge or the failure to use general problem-solving methods will result in failure to solve problems.

In a similar vein, consider the following logarithmic equation:

$$\log_{10}(x - 1) + \log_{10} 5x = 1$$

Stop reading and try to solve the above equation for the value of the variable x.

The solution is analogous to the solution of the previous problem; namely, we set as a subgoal the derivation of an equation that is a simple polynomial in x, for which we may know a solution method (for example, factoring or substitution into the quadratic formula). In this instance, we set as an initial subgoal the derivation of an equation involving no log terms. That is, we attempt to eliminate logs in the above equation. Stop reading and try to solve the problem, making use of this subgoal.

Eliminating logarithms from the equation can be achieved by exponentiating each side of the equation, as follows:

$$10^{\log(x-1)+\log 5x} = 10^1$$

$$10^{\log(x-1)} \cdot 10^{\log 5x} = 10$$

$$(x-1) \cdot 5x = 10$$

$$5x^2 - 5x - 10 = 0$$

$$x^2 - x - 2 = 0$$

$$(x-2)(x+1) = 0$$

$$x = 2$$

or

$$x = -1$$

From the above, two roots for the equation were derived: $x = 2$ and $x = -1$. The latter root is not a solution to the original equation; it is a root that was added via the exponentiating process, since exponentiating increased the degree of the equation. Operations that result in equations with added roots are not as dangerous to use as operations that result in equations that eliminate roots of the original equation. When roots are added, it is easy enough to determine the correct roots by substitution in the original equation. When roots are subtracted, there may be no way to determine the value of the eliminated root — which may not be a serious problem, if you do not need to get that eliminated root.

In the present problem, exponentiating both sides of the equation resulted in a quadratic that was factorable, permitting easy solution. Of course, any quadratic equation can be solved by the quadratic formula, which anyone who has mastered high school algebra should have memorized or be able to look up. In this and the preceding problem, the state evaluation function being used was that equations with logs or exponentials of polynomials in x are less highly valued than equations that are simple polynomials in x, regardless of the degree. The reason is that we do not know any direct algorithm for solving an exponential or logarithmic equation. Thus, we are required to transform the equation into some form for which we know a method of solution that works at least in some cases. In the last two problems, then, we had to transform the exponential or logarithmic equations into polynomial equations, hoping that the polynomial equations so derived would be solvable by factoring or substitution into the quadratic formula.

TRIGONOMETRY

Determine the altitude, h, of a general scalene triangle, given the length of one side (its base b) and the angles made by the two other sides with the base (the two base angles, α and γ), as illustrated in Fig. 11-1.

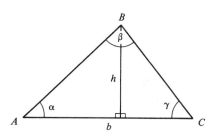

FIGURE 11-1
Altitude of a triangle problem.

Knowing the base b and the angles α and γ, we have determined a specific (unique) triangle. Thus, in principle, the altitude and every other property of the triangle is specified. However, to compute h, we need to have relationships that link h to the values of the known quantities α, γ, and b. We might consult a trigonometry text to determine whether there was any simple formula involving the unknown h and the three known quantities α, γ, and b. Suppose the trigonometry text listed no such formula. How might we proceed to determine the value of h? Stop reading and try to solve the problem.

We might define a subgoal of determining the area of the triangle in terms of the known quantities α, γ, and b. This subgoal is extremely useful since we already should know that an equation exists that relates h and b to the area (A) of the triangle—namely, $A = \frac{1}{2}bh$. Stop reading and try again to solve the problem, if you did not do so before.

Suppose the trigonometry text does indeed list several formulas for the area of a triangle, one of which looks somewhat similar to the equation we are after, namely, $A = (b^2 \sin \alpha \sin \gamma)/(2 \sin \beta)$. The only problem with this equation is that it involves the additional quantity β. However, if we remember that the sum of the angles of a triangle equals 180°, then knowing two angles of a triangle allows us to compute the value of the third angle of the triangle, namely, $\beta = 180 - \alpha - \gamma$. Thus, using this equation for the area of a triangle in terms of the three known quantities α, γ, and b, we can determine the area, A. From knowledge of the area, A, and the base, b, we can compute the height, h, which was to be determined.

Once again, note that specific knowledge of trigonometric and geometric relations is critical for solving the problem. However, the more complex formula for the area of the triangle in terms of three angles and one side is a relation that we need not have memorized but only be capable of looking up in a trigonometry text. Even the simple formula for finding the area of the triangle in terms of its height and its base might be looked up in such a text, though it is likely to be remembered by a student who has understood geometry and trigonometry. The specific geometric knowledge that the sum of the angles of the triangle is 180° probably must be known in order to solve this problem.

Solving this problem requires more than knowledge of the relevant geometric and trigonometric facts. We must also know which of all of the relevant facts should be selected to use in the solution. This selection requires the use of general problem-solving methods. The goal is to find the value of h. We might work backward from this statement of the goal to an equation in which h is involved along with some other quantities. Such an equation might be $A = \frac{1}{2}bh$, the most commonly known formula for the area of a triangle. Since this equation involves the area that is also unknown, it immediately suggests the subgoal of finding the value of the area in terms of known quantities. This subgoal probably requires us to examine a table of trigonometric formulas to determine if there is a formula that relates the known quantities α, γ, and b to the area of the triangle. If there is, then we have all we need to solve the problem. For us to solve this problem, having a specific knowledge of trigonometric formulas is less important than having access to books containing such information in conveniently usable form. What is critical is the use of the general problem-solving methods of working backward and defining subgoals. These methods provide the framework within which we can proceed in a goal-directed manner to solve the problem.

Hill climbing was also used to some extent to solve this trigonometry problem. In working backward, the selection of an equation relating h to b and A is superior to an equation relating h to quantities none of which is known. Here, the value of the evaluation function is greater, the more known quantities appear in the expression and the fewer unknown quantities appear in the expression. The same principle applies in working forward in trying to define as a subgoal another formula for the area of the triangle that involves as many known quantities as possible. In this case, it was possible to find a formula that involved all three known quantities and an additional quantity that was not originally specified in the givens as known but that could be trivially derived from the given information by means of a well-known geometric theorem (that the sum of the angles of a triangle is 180°).

ANALYTIC GEOMETRY

Determine the location and geometric properties of the figure specified by the equation $x^2 + y^2 - 5x + 7y = 3$.

The specifically relevant knowledge from the field of analytic geometry is that any equation of the form $x^2 + y^2 + Ax + By = C$ is the equation of a circle, and this equation can always be transformed to an equivalent equation of the form $(x - a)^2 + (y - b)^2 = c^2$, where (a, b) represents the coordinates of the center of the circle and c represents the radius of the circle. Retrieving this relevant knowledge is surely essential to solving the problem, and it gives part of the answer already—namely, that the form of the figure specified by the above equation is the form for a circle. All that remains is to determine the coordinants of the center of the circle and its radius (a, b, and c). It is this problem to which general problem-solving methods are applicable. Stop reading and try to solve the problem.

The given expression is $x^2 + y^2 - 5x + 7y = 3$, and the goal expression is of the form $(x - a)^2 + (y - b)^2 = c^2$. Once again, we might define a subgoal by means of working backward from the goal expression. Starting with the goal expression, we know we can rewrite the goal expression in the form $(x^2 - 2ax + a^2) + (y^2 - 2by + b^2) = c^2$. Now we need to work forward from the given expression to the subgoal expression. Stop reading and try again to solve the problem, if you did not do so before.

We know that $2a = 5$, from which $a = \frac{5}{2}$. Similarly, $-2b = 7$ or $b = -\frac{7}{2}$. This means that we must add $a^2 = 25/4$ and $b^2 = 49/4$ to the left side of the given equation in order to complete squares to get to the left-hand side of the subgoal equation. The same $29/4 + 49/4$ must, therefore, be added to the right side of the given equation, yielding $3 + 74/4$, or $86/4$. Thus, $86/4 = c^2$ and $c = \sqrt{86}/2$. Therefore, the coordinates of the center of the circle are $(\frac{5}{2}, -\frac{7}{2})$, and the radius of the circle is $\sqrt{86}/2$, and the problem is solved.

Completing the square becomes a rather familiar specific technique in and of itself, once you have a certain degree of experience in mathematics. However, at some point to the beginning mathematics student, it is a new, unknown technique. To the same student, the technique of expanding a term of the form $(x-a)^2$ to get $(x^2 - 2ax + a^2)$ is familiar. By using the general problem-solving method of working backward to get a subgoal, we can get the idea of completing the square in a completely natural way and know exactly how to do it.

> Determine equations for the new coordinates of a point in a plane when the new coordinate system is obtained by translation and rotation from an old coordinate system. Translation of a coordinate system means the origin is changed to a new point, and rotation of a coordinate system means both axes are turned through the same angle in the same direction, pivoting about the origin.

It makes no difference in which order the translation and rotation operations are performed; the same new coordinate system is obtained in either case. Stop reading and try to solve the problem.

The solution of this problem results simply from breaking the problem into parts, that is, setting subgoals. First, solve the problem of characterizing the new coordinates obtained by simple translation alone. Having achieved this subgoal, then solve the second subgoal of characterizing the final set of coordinates after a rotation has been applied to the coordinate system previously derived from the translation. Stop reading and try to solve the problem again.

If x and y are the original coordinates, and the coordinates of the new origin in terms of the original (x, y) axes are (x_0, y_0), then let the new coordinates obtained by translation be represented by x_1, y_1. The formulas for simple translation of coordinates are as follows:

$$x = x_1 + x_0 \qquad x_1 = x - x_0$$
$$y = y_1 + y_0 \qquad y_1 = y - y_0$$

Stop reading and try to solve the rest of the problem, if you have not done so already.

Let the coordinates obtained by rotation through the angle α be x_2 and y_2. Formulas relating the new (x_2, y_2) coordinates to the previous (x_1, y_1) coordinates are as follows:

$$x_1 = x_2 \cos \alpha - y_2 \sin \alpha$$

$$y_1 = x_2 \sin \alpha + y_2 \cos \alpha$$

To get formulas for the combined translation and rotation transformation, simply combine the above formulas to obtain the following:

$$x = x_2 \cos \alpha - y_2 \sin \alpha + x_0$$

$$y = x_2 \sin \alpha + y_2 \cos \alpha + y_0$$

To express the new coordinates in terms of the old coordinates requires some algebra, from which yields the following:

$$x_2 = x \cos \alpha + y \sin \alpha - (x_0 \cos \alpha + y_0 \sin \alpha)$$

$$y_2 = -x \sin \alpha + y \cos \alpha - (-x_0 \sin \alpha + y_0 \cos \alpha)$$

With the problem being broken into two subgoals, each of which was simpler to obtain, it was possible to obtain the solution to the original problem by simple algebraic combination of the solutions to the two subproblems.

Now let us consider the following problem involving the transformation of coordinate systems:

Is it possible to transform axes such that the straight line $4x - 3y + 2 = 0$ will have the form $x_1 = 0$ and such that the straight line $2x + y = 4$ will have the form $x_1 = ay_1$? If such is possible, derive the transformation.

Stop reading and try to solve the problem.

Probably the first thing to note is that we have two equations and two unknowns. If the equations are independent, which they are, then it will be possible to solve the equations for the value of x and y. From a geometric point of view, then, you must find the point of intersection of the two straight lines represented by these linear equations. Solving for the point of intersection of the two straight lines will prove to be an important subgoal in solving the problem, but you need not even realize that in the beginning. Since solving these two linear equations and two unknowns is so simple, you should probably simply draw these inferences from the given information, without regard to the goal, at the outset of the problem (as discussed in Chapter 3). This assumes, of course, that you are familiar with the process of solving two linear equations with two unknowns, so that this is a trivial inference. When it is easy to represent explicitly the information that is given implicitly in the problem, you should undoubtedly do so in the beginning, before even thinking about how to reach the goal from the given information. This initial step will yield the information that the solution of the two equations (point of intersection of the two straight lines) is $x = 1, y = 2$. Stop reading and try to solve the problem, if you did not before.

Consider the goal and derive a suitable plan for achieving the goal from the given information. Having drawn the inference that the two straight lines intersect at a particular known point, it is now clear that the goal is achievable. Why? Stop reading and try to solve the problem, if you still have not done so.

The problem obviously indicates a division into two subproblems: (a) making the first line have the form $x_1 = 0$ and (b) making the second line have the form $x_1 = ay_1$. Drawing inferences about these subgoals will point out the general types of transformation necessary to achieve each subgoal. Stop reading and try to solve the problem, if you did not do so.

It is somewhat easier to consider the achievement of the second subgoal first. The second subgoal equation, $x_1 = ay_1$, asserts that, in the new coordinate system, the second line will have zero x_1 and y_1 intercepts (being a strict proportionality). Having zero x_1 and y_1 intercepts means that the equation of the second line must pass through the origin of the coordinate system. Having drawn this inference from the subgoal (or worked backward from the subgoal, if you will), what transformation of the original coordinate system will achieve this new subgoal? Stop reading and try to solve the problem, if you did not before.

Clearly, we can achieve the second subgoal by a simple translation of the coordinate system from the origin to any point on the second line $(2x + y = 4)$. Our preceding inference concerning the point of intersection of the two straight lines might bias us to translate the origin of the coordinate system to the point of intersection of the two straight lines, but at this stage of working on the problem we do not know for sure that this is the correct point of origin for the new coordinate system.

Now, how do we achieve the other subgoal, namely, that of transforming the first line so that it has the form $x_1 = 0$? Stop reading and try to solve the problem, if you have not done so already.

Again drawing inferences from the subgoal (working backward), we see that to transform the first line into a line with the equation $x_1 = 0$, we must transform the first line so that it is coincident with the y axis in the new coordinate system. Such a line will always have the x_1 coordinate equal to zero for any value of the y_1 coordinate. Achieving this goal requires what type of transformation of the original coordinate system? Stop reading and try to complete the solution of the problem, if you have not done so already.

Clearly, to achieve this subgoal we must, first, translate the origin of the coordinate system to some point along the first line and, second, rotate the axis of the coordinate system to coincide with the first line. The first aspect interacts with the transformation necessary to achieve the other subgoal, so we are restricted to locating the origin of the new coordinate system at the point of intersection of the two straight lines (since the origin of the new coordinate system must lie on both straight lines). A second restriction in achieving the present subgoal is that the axis of the coordinate system must be rotated around this new origin until the y axis coincides with the first straight line $(4x - 3y + 2 = 0)$. Geometric intuition indicates that this can be done and so achieving the goal is clearly possible.

Now the problem is to derive the nature of the rotation, since the translation is already obvious, that is, to move the origin to the point $(1, 2)$. To solve the rest of the problem, we now need to know the angle of rotation of the coordinate system required to line up the y axis with the straight line represented by $4x - 3y + 2 = 0$. Stop reading and solve the rest of the problem, if you have not done so already.

To solve this subproblem let us write down the formulas for the new coordinates in terms of the old coordinates (noting that $x_1 = x_2$ and $y_1 = y_2$) – namely, $x_1 = x \cos \alpha + y \sin \alpha - k$, where $k = x_0 \cos \alpha + y_0 \sin \alpha$. We might also go ahead and write down the equation for y_1 in terms of x, but this actually is unnecessary to the solution of the problem. If we are continually aware of what terms represent constants and what terms represent variables in any expression, then we should note that, in the right side of the equation we have just written down, x and y are the only variables, and $\cos \alpha$, $\sin \alpha$, and k are all constants (albeit unknown to us at present). Since given information specifies that $x_1 = 0$ and that $4x - 3y + 2 = 0$, we can equate $x \cos \alpha + y \sin \alpha - k$ to the expression $4x - 3y + 2$.

Now if we take the equations for the general transformation of coordinates involving both translation and rotation and substitute them into the equation $4x - 3y + 2 = 0$ we obtain

$$4(x_1 \cos \alpha - y_1 \sin \alpha + x_0) - 3(x_1 \sin \alpha + y_1 \cos \alpha + y_0) + 2 = 0.$$

Since we wish to find an α for which $x_1 = 0$, we can substitute $x_1 = 0$ into the above equation and also substitute the known values of $x_0 = 1$ and $y_0 = 2$. This yields the equation

$$-(4 \sin \alpha + 3 \cos \alpha)y_1 = 0$$

This equation implies that $4 \sin \alpha + 3 \cos \alpha = 0$. Thus, $4 \sin \alpha = -3 \cos \alpha$. To determine α from this equation, simply remember the trigonometric identity that $\sin \alpha/\cos \alpha = \tan \alpha$. We can then derive from the above that $\tan \alpha = -\frac{3}{4} = -0.75$. Using the tables, this indicates that $\alpha = -36° \, 52'$. Thus, the solution to the problem is to translate the origin of the coordinate system to the point $(1, 2)$ and rotate the coordinate system through an angle of $36° \, 52'$ in the negative direction.

One slightly tricky aspect of the problem is the use of the equation $x_1 = 0$ in conjunction with the equation derived by substituting into $4x - 3y + 2 = 0$. After we have established that it is indeed possible to derive a transformation of the coordinates so that the equation $4x - 3y + 2 = 0$ can be transformed into an equation of the form $x_1 = 0$, we are not really trying to achieve the goal expression $x_1 = 0$. Rather $x_1 = 0$ is part of the given information that we are using to achieve the subgoal of determining the angle of rotation α. This problem clearly points up the need to carefully define and redefine what is given information and what is the goal at different stages in the solution of the problem.

CALCULUS

> Prove that, within the set of triangles having a constant base and constant perimeter, the isosceles triangle has the maximum area.

The specific calculus necessary to solve this problem is to know that we can often find the maximum or minimum of a function by differentiating it with respect to the variable(s) of which it is a function. Stop reading and try to solve the problem.

Clearly, in the present instance, the function (dependent variable) is the area of a triangle. However, the area of a triangle can be expressed as a function of a number of different independent variables. Therefore, let us set the subgoal of finding a formula for the area of a triangle that involves those quantities that are specifically given in this problem (either constants or the variables with respect to which we wish to find maximum area). Stop reading and try again to solve the problem, if you did not do so before.

In this problem, the constants and independent variables are evidently the sides of the triangle (including the sum of the sides, which is the perimeter). Thus, we need to find a formula that involves only these quantities. Such a formula, which can be looked up in a book, is Heron's Formula: $A = [s(s - a)(s - b)(s - c)]^{1/2}$, where A is the area of a triangle; s is the semiperimeter, which equals $\frac{1}{2}(a + b + c)$; and a, b, and c are the lengths of the sides. Stop reading and try to solve the problem, if you have not done so already.

Having achieved the first subgoal of finding a formula that involves the relevant constants and variables, we should note that the formula contains the semiperimeter (which is a constant, since the perimeter is a constant) and the length of one side (which is a constant; let it be side a). Two variable sides remain – namely, b and c. Suppose we are familiar with differentiating functions of single variables with respect to the single variable in order to obtain a maximum or minimum of the function with respect to that variable. To use this knowledge, we set a second subgoal to reduce the number of independent variables from two to one. Stop reading and try to solve the problem if you have not already done so.

We can achieve the second subgoal, since with a constant perimeter and a constant base (a), the sum of the other two sides must be equal to a constant. Thus, $b + c = k$ and $c = k - b$. Substituting $c = k - b$ into Heron's Formula for the area of a triangle, we obtain the area as a function of a single variable, namely, the length of side b. This can be differentiated with respect to b and the derivative set equal to zero to determine that $k = 2b$. A trick that makes this a bit easier is to note that, if the area A is a maximum, then A^2 is a maximum and vice-versa. Since it is somewhat easier to differentiate A^2 with respect to b than to differentiate A with respect to b, this little trick saves some work. In either case, we solve for $k = 2b$, from which it follows that $b = c$, and the theorem is proved. The work is given below:

$$A^2 = s(s - a)(s - b)(s - c)$$

$$A^2 = s(s - a)(s - b)(s - k + b)$$

$$dA^2/db = s(s - a)[(-1)(s - k + b) + (s - b)(+1)] = 0$$

$$-s + k - b + s - b = 0$$

$$k - 2b = 0$$

$$k = 2b$$

$$b + c = 2b$$

$$c = b$$

Incidentally, the above problem can be solved entirely without calculus, using the method of contradiction in conjunction with Heron's Formula. To use the method of contradiction, we assume that the squared area (A_2^2) in the case where $b_2 \neq c_2$ is greater than the squared area (A_1^2) in the case where $b_1 = c_1 = d$. Without loss of generality, assume $b_2 > c_2$. Since $b_2 + c_2 = 2d$, then $b_2 > d > c_2$. Using these equalities and inequalities and some algebraic manipulation of the equation $(s - b_2)(s - c_2) > (s - d)^2$, we can eventually derive $(b_2 - c_2)^2/4 < 0$ which is a contradiction since the square of any real number must be positive. The algebra is given below for the interested reader:

$$A_2^2 \overset{?}{>} A_1^2$$
$$s(s - a)(s - b_2)(s - c_2) \overset{?}{>} s(s - a)(s - d)(s - d)$$
$$(s - b_2)(s - c_2) \overset{?}{>} (s - d)^2$$
$$b_2 = 2d - c_2$$
$$(s - 2d + c_2)(s - c_2) \overset{?}{>} (s - d)^2$$
$$s^2 - sc_2 - 2ds + 2dc_2 + sc_2 - c_2^2 \overset{?}{>} s^2 - 2ds + d^2$$
$$2dc_2 - c_2^2 \overset{?}{>} d^2$$
$$c_2(2d - c_2) \overset{?}{>} d^2$$
$$c_2 b_2 \overset{?}{>} \left(\frac{b_2 + c_2}{2}\right)^2$$
$$b_2 c_2 \overset{?}{>} \frac{b_2^2}{4} + \frac{b_2 c_2}{2} + \frac{c_2^2}{4}$$
$$0 \overset{?}{>} \frac{b_2^2}{4} - \frac{b_2 c_2}{2} + \frac{c_2^2}{4}$$
$$0 \overset{?}{>} \left(\frac{b_2 - c_2}{2}\right)^2$$

Derive the form of the following indefinite integral:

$$y = \int \frac{1}{1 + x^{1/2}} \, dx$$

The background information that is assumed to be given includes knowledge of the integrals of elementary functions (such as x^n, e^x, $\log x$, $\sin x$, and $\cos x$). Other important background information are the techniques of integration by substitution and by parts, and differentiating a function of a function.

The first major choice in attacking an integration problem of this kind is whether to use the method of substitution or the method of integration by parts. Sometimes both methods must be used, but, in any event, you still have to decide which to apply first. Since integration by substitution is the more useful technique, it is to be preferred as an initial choice of integration method, unless there is some special reason for preferring integration by parts. Integration by parts is useful primarily when the function to be integrated is a product of two functions $[f(x) = g(x)h(x)]$. Although all functions of x can be written as a product of two functions—namely, $f(x) = f(x) \cdot 1$—this is a trivial type of product to which the application of integration by parts is only occasionally useful. Thus, in the present problem, there is no reason to use integration by parts, so we adopt integration by substitution as our initial operation. Stop reading and try to solve the problem.

Note that the problem-solving method considerations discussed so far in this problem are all specific to calculus. Substitution and integration by parts are not general problem-solving methods. However, in deciding what type of substitution to make, general problem-solving methods play some role. In particular, hill climbing is useful. The hill climbing uses an evaluation function concerned roughly with simplicity of functional form and the likelihood of your knowing an integral for the function resulting from this substitution. At the same time, another evaluation function is generally working at cross purposes with the first one—namely, the simplicity of the functional form for the substitution $u = g(x)$. A good rule of thumb is to try a substitution whose functional form is less complicated than that of the original function and results in a function to be integrated that is also less complicated than the original function. Thus, in the present problem, although a substitution of the form $u = (1 + x^{1/2})^{-1}$ would greatly simplify the original problem, the form of the substitution function would be as complicated as the original function to be integrated. Stop reading and try again to solve the problem, if you did not before.

Better substitutions would be $u = x^{1/2}$ or $u = 1 + x^{1/2}$. Using the latter substitution, $u = 1 + x^{1/2}$ and $dx = 2(u - 1)\ du$, yields

$$\int \frac{1}{1 + x^{1/2}}\ dx = 2 \int \frac{(u-1)}{u}\ du = 2 \int 1 - \frac{1}{u}\ du = 2(u - \log u) + c$$

$$= 2(1 + x^{1/2}) - 2 \log (1 + x^{1/2}) + c$$

$$= 2x^{1/2} - 2 \log (1 + x^{1/2}) + K$$

Frequently, several substitutions will be required in order to solve the problem, and, at each step you are essentially using the method of hill climbing on an evaluation function concerned with simplicity of functional form. There is no precise definition of simplicity of functional form, but that lack should not prevent you from explicitly recognizing that this is what you are doing and that you have rather good judgment as to what functions are simpler than other functions (in the sense of being closer to functions for which you know the integral). As long as you are able to decide that the functions resulting from certain substitutions are simpler than functions resulting from other substitutions, you are in a position to make good use of the hill-climbing method, whether or not you can explicitly define the evaluation function.

Derive the functional form of the following integral: $\int x^2 e^x \, dx$. The specific background knowledge includes knowledge of the integrals of the elementary functions plus the integration by parts formula—namely, $\int uv \, dx = uV - \int u'V \, dx$, where $V = \int v \, dx$ and $u' = du/dx$.

Since the function to be integrated is an obvious product of two simpler functions, the method of integration by parts is suggested. Whether integration by parts is making progress toward the goal is determined a great deal by the general problem-solving method of hill climbing on an evaluation function of simplicity of functional form and ease of integration. Stop reading and try to solve the problem.

In the present case, two applications of the method of integration by parts is necessary in order to solve the problem. At each stage the application of integration by parts results in functions to be integrated that are simpler than the functions to be integrated prior to the application of integration by parts. The specific solution is as follows:

$$\int x^2 e^x \, dx = x^2 e^x - \int 2xe^x \, dx + K$$
$$= x^2 e^x - 2xe^x + 2e^x + C$$
$$= (x^2 - 2x + 2)e^x + C$$

In the present instance, it would be possible to give a precise definition of the evaluation function on which the hill climbing is occurring—namely, the exponent of x in the product $x^n e^x$ when this product is the function to be integrated. Repeated application of the method of integration by parts results in reduction of the exponent, eventually to $x^0 e^x$ or e^x. However, whether or not it is possible to explicitly define the evaluation function being used, hill-climbing methods can be extremely useful in solving such a problem, so long as your judgment of simplicity is reasonably accurate.

Find the values of x for which the function $y = f(x)$ is a maximum or minimum. The function is defined by the equation $x^2 + xy + y^2 = 27$. Relevant background information includes the chain rule for differentiating the function of a function, the rule for differentiating the product of two functions, and the theorem that the derivative of a function equals zero at a minimum or maximum.

When we are finding the maxima or minima of even a function of a single variable, $y = f(x)$, we are essentially solving two equations for the values of two unknowns, x and y. This fact is often not apparent to students when they originally learn the method of finding maxima and minima by differentiating $f(x)$, setting it equal to zero, and solving for x, because the original equation was already solved for y as a function of x. In such a case, the derivative will involve only a single variable x. When the derivative is set equal to zero, the resulting equation is solved for the value of x for which the function is a maximum or minimum. In the present problem, the initial function $y = f(x)$ is defined implicitly by the equation $x^2 + xy + y^2 = 27$. In this case, it is necessary to take a more general approach to the problem, in which finding the derivative and setting it equal to zero allows us to obtain a second equation, in addition to the equation $x^2 + xy + y^2 = 27$. We hope that these two equations will permit us to solve for the (x, y) points for which the function has a maximum or minimum. Stop reading and try to solve the problem.

To solve the problem, we should initially set a subgoal: to obtain an equation that involves the derivative $y' = dy/dx$. Stop reading and try again to solve the problem, if you did not before.

This subgoal can be achieved by differentiating the given equation with respect to x (employing the product and chain rules for differentiation). The resulting equation is $2x + xy' + y + 2yy' = 0$. This equation can be solved for y' by algebraic manipulation, yielding the equation

$$y' = \frac{-(2x + y)}{x + 2y}$$

When this equation is set equal to zero, we obtain $y = -2x$. Substituting this equation into the original equation, we obtain $x^2 - 2x^2 + 4x^2 = 27$ or $3x^2 = 27$ or $x = \pm 3$, $y = \pm 6$, and the problem is solved. Once again, a simple definition of a single subgoal—namely, obtaining an expression for y' in terms of x and y—resulted in straightforward solution of the problem.

DIFFERENTIAL EQUATIONS

The solutions of differential equations provide particularly good examples of the use of the problem-solving methods of hill climbing, subgoals, and analogy to similar problems. Perhaps the most important specific training is the ability to place a differential equation in the proper class. Once you note what other differential equations the one in front of you is similar to (what class it belongs to), you can then apply the techniques associated with the solution of that class of differential equations. You need not even have much specific knowledge of how to solve equations of a particular class, so long as you can identify the class and look up in a book how to solve equations of that class. Thus, analogy to similar problems is the crucial first step in the solution of many differential equations.

When the given differential equation is a member of a class for which solution methods are known, the methods of hill climbing and subgoals (using evaluation functions) are also quite important. For example, in solving differential equations of different forms, we often proceed by setting as a subgoal the transformation of the differential equation into another differential equation of simpler form, and then by using the known solution methods for the simpler form.

In grab-bag classes of differential equations (such as miscellaneous nonlinear differential equations), we may attempt to define subgoals such as transforming the equation to linear form or reducing the order or degree of the equation, but frequently we simply try out various operations on the given nonlinear differential equations to see which ones result in an equation of the simplest form. The latter is clearly an example of hill climbing, using an evaluation function that somehow weights different features of a differential equation for overall ease of solution.

For equations with order m and degree $n(m, n > 1)$, the relevant evaluation function is frequently the vector (m, n), with lower values of either m or n being more highly valued. Differential equations provide good examples of vector evaluation functions, where there are many different properties on which hill climbing might be tried to see which, if any, approach would solve the problem. Frequently, the solution of a differential equation requires a sequence of steps in which the degree and order of the equation are progressively reduced, finally resulting in a differential equation of the first order and first degree. The order in which the degree and order of the differential equation is progressively reduced may vary from problem to problem.

Once the stage is reached where you have a nonlinear differential equation of the first order and first degree to solve, a variety of potential solution sequences can follow, again depending on the type of first-order, first-degree differential equation.

The nonlinear differential equation may be reducible to linear form by some suitable transformation. No general rules exist for determining such transformations nor the types of nonlinear equations to which they apply, but experience with a wide variety of such problems may indicate that the present problem is similar to some problem already solved in this way. Having achieved a linear equation of the first order and first degree, you then apply solution methods appropriate to this type of equation (for example, using Laplace transforms or integrating factors).

Another solution sequence starting with a nonlinear, first-order, first-degree differential equation (which is appropriate in some cases) is to try to transform the equation to be a member of a particular class of differential equations known as *exact differential equations*. Achieving this subgoal may require you to find an appropriate integrating factor to transform the given differential equation into an exact differential equation. Once an exact differential equation has been obtained, you simply follow solution methods appropriate for this type of equation.

Another solution sequence is appropriate to differential equations of the form

$$\frac{dy}{dx} = \frac{a_1 x + b_1 y + c_1}{a_2 x + b_2 y + c_2}$$

where $a_1 b_2 - a_2 b_1 \neq 0$. To solve such equations, we set the subgoal of transforming this inhomogeneous equation into a homogeneous equation by making a substitution. The next subgoal is to transform this homogeneous equation into a differential equation in which the variables are separated, which is then solved by direct integration.

There are other solution sequences appropriate to other types of nonlinear differential equations of the first order and first degree. However, just considering the solution sequences discussed here, note that an experienced solver of differential equations has established an evaluation function for first-order, first-degree differential equations, which is essentially a partial ordering of a variety of different forms of such differential equations. In this partial ordering, equations with separated variables are more highly valued than homogeneous equations in which the variables are not separated, the latter being more highly valued than the type of inhomogeneous equations described above, which in turn are more highly valued than many nonlinear differential equations not of this or any other identifiable type. At the same time, exact differential equations are more highly valued than these miscellaneous nonlinear differential equations, but there is no relative ordering of exact differential equations relative to equations within another solution sequence, such as that appropriate for the inhomogeneous equations of the previously specified type. Along the same lines, linear differential equations are more highly valued than the miscellaneous nonlinear differential equations, but they are not necessarily more highly valued than types of equations within some other solution sequence. This is what we mean by saying that these different types of first-order, first-degree differential equations have an evaluation function in the form of a *partial ordering*, rather than in a complete or simple rank ordering of all the different types of such equations.

If we know a variety of such types of differential equations and the appropriate partial ordering type of evaluation function defined over them (that is, know the variety of different solution sequences), we are in a good position either to define subgoals or to recognize progress in the use of the hill-climbing method. However, there is still the problem of determining the proper operation (substitution, integrating factor, and the like) to take in order to achieve a differential equation of the more highly evaluated (simpler) form. Perhaps general problem-solving methods are applicable to this aspect of solving differential equations, but, frankly, I have so little experience in solving differential equations that I feel incompetent to discuss the matter further.

In any event, once again, a solution of mathematical problems requires a mixture of specific knowledge of mathematics and the use of general problem-solving methods. You can, of course, learn how to solve differential equations and other mathematical problems without appreciating that you are thereby applying general problem-solving methods. However, understanding general problem-solving methods probably facilitates your understanding the variety of techniques applicable to such mathematical problems. A generally accepted dogma in educational psychology is that the more you can relate new knowledge to old knowledge, the faster and more complete your learning will be (though how good the evidence for this is I certainly do not know). So, if you know general problem-solving methods you should be able to quickly organize many specific methods for solving differential equations when these methods are introduced in terms of defining classes of similar problems and defining evaluation functions that permit you to use the subgoal and hill-climbing methods.

To complement this rather abstract discussion, let us consider the solution of the following differential equation, which was produced by Al Stevens, a student in one of my problem-solving classes:

$$2y^2 \frac{d^2y}{dx^2} + 2y\left(\frac{dy}{dx}\right)^2 = 1$$

Stop reading and try to solve this differential equation.

Stevens first defined as a subgoal the transformation of this equation into a linear, second-order differential equation, but he quickly replaced this subgoal with a different subgoal—namely, that of reducing the equation from second order, nonlinear to first order, nonlinear. Stop reading and try again to solve the problem, if you did not do so before.

The second subgoal is easily achieved because the differential equation is of the form $d^2y/dx^2 = f(y, dy/dx)$, with d^2y/dx^2 not being a function of x. By recognizing the differential equation as a member of this class, Stevens made available his knowledge that a standard substitution—namely, $v = dy/dx$—would transform the second-order differential equation into a first-order differential equation.

Since $dv/dx = d^2y/dx^2$ and $dv/dx = (dv/dy)(dy/dx)$, then $d^2y/dx^2 = v\, dv/dy$. Substituting into the original equation yields the first-order nonlinear differential equation $2y^2v\, dv/dy + 2yv^2 = 1$. Stop reading and try again to solve the problem, if you did not do so before.

Algebraic manipulation of the equation yields $dv/dy + y^{-1}v = \frac{1}{2}y^{-2}v^{-1}$. Such an equation belongs to another specific class of differential equations—namely, Bernoulli equations—which are of the form

$$\frac{dv}{dy} + P(y)v = Q(y)v^n$$

You can look up in a book that such equations can be reduced to linearity by the substitution $w = v^{1-n}$. In this case, that means the substitution $w = v^2$, for which

$$v = w^{1/2} \quad \text{and} \quad \frac{dv}{dy} = \frac{1}{2}w^{-1/2}\frac{dw}{dy}$$

Substitution then yields

$$\frac{1}{2}w^{-1/2}\frac{dw}{dy} + y^{-1}w^{1/2} = \frac{1}{2}y^{-2}w^{-1/2}$$

Multiplying through by $2w^{1/2}$ yields

$$\frac{dw}{dy} + 2y^{-1}w = y^{-2}$$

which is a linear equation in w and y. From this point, the solution is straightforward by methods that can be looked up in a book.

In retrospect, Stevens noticed that his first-order, nonlinear differential equation, $2y^2v\, dv/dy + 2yv^2 = 1$, had the form $d(y^2v^2)/dy = 1$, in which the variables y and y^2v^2 are trivially separable. Thus, the equation is easily solvable without it having to be reduced to linear form. The solution is as follows:

$$d(y^2v^2) = dy$$

Integrating,

$$y^2v^2 = y + c_1$$

Since $v = dy/dx$,

$$y^2\left(\frac{dy}{dx}\right)^2 = y + c_1$$

$$y\,\frac{dy}{dx} = (y + c_1)^{1/2}$$

$$\int \frac{y}{(y + c_1)^{1/2}}\,dy = \int dx$$

From a table of integrals,

$$\frac{-2(2c_1 - y)\,(c_1 + y)^{1/2}}{3} = x + c_2$$

PROBABILITY AND STATISTICS

A sample of two observations, x and y, are drawn from the uniform distribution on the interval from zero to 1, $f(x) = f(y) = 1$, $0 \le x, y \le 1$. Find the rth raw moment of z, $(\mu_{r:z})$, where $z = xy$. Note that

$$\mu_{r:z} = \int_{-\infty}^{\infty} z^r g(z)\,dz$$

There are at least three different ways of solving this problem. Stop reading and try to think of as many ways as you can.

The most obvious (but also the most difficult) way is to set two subgoals: (a) that of finding the probability density function $[g(z)]$ for the new random variable $z = x \cdot y$ and (b) that of plugging this probability density function into the definition for the rth raw moment of z.

Let us assume that we know a simple generalization of the above formula that extends it in order to find the rth moment of a *function* of random variables, $h(x, y)$, where the joint probability density function for the random variables x and y is represented by $f(x, y)$. The formula is

$$\mu_{r:h(x,y)} = \int_{-\infty}^{\infty} [h(x,y)]^r f(x,y) \, dx \, dy$$

Now stop reading and try to solve the problem, if you did not do so before.

In this case, we can compute the rth moment of the function $z = x \cdot y$, provided we know the probability density function of the joint distribution of x and y. Thus, the first subgoal is to determine this function, $f(x, y)$. We must assume as implicit information (though it was not specifically stated in the problem) that the two sample observations (x and y) are independent. Knowing these observations are independent and knowing the density functions for each, we get the joint density function $f(x, y) = f(x)f(y) = 1$, $0 \le x, y \le 1$, $f(x, y) = 0$, elsewhere. From this point on, the solution is a simple integration, as follows:

$$\mu_{r:z} = \mu_{r:h(x,y)} = \int_0^1 (xy)^r f(x,y) \, dx \, dy$$

$$= \int_0^1 x^r \, dx \cdot \int_0^1 y^r \, dy$$

$$= \frac{1}{(r+1)^2}$$

Finally, one can set a totally different subgoal of finding the moment generating function for the new variable $z = xy$ and differentiating the moment generating function r times to find the rth moment of z. In solving the problem by this method, we need to know more specific background information (such as the definition of a moment generating function and the Taylor series expansion for $e^{\theta x}$), but otherwise the problem is solved in a straightforward manner by this method as well.

The principal general problem-solving method used to solve this problem was the setting of subgoals. A variety of such subgoals were logically related to the solution of the problem—namely, that of deriving the probability density function $g(z)$, that of deriving the joint distribution function $f(x, y)$, or that of deriving the moment generating function for z. Setting the subgoals in each case is a part of an overall calculative plan for solving the problem in each of the three cases. For example, a first step in solving the problem might be to write down the formula for the rth moment of z in terms of a double integral of $(xy)^r$ and the joint probability density function $f(x, y)$. This would suggest that we needed to determine the joint probability density function as a subgoal in order to do the integration and solve the problem.

> The lengths of two parts, A and B, are normally distributed with means $\mu_A = 2$ centimeters and $\mu_B = 4$ centimeters and standard deviations $\sigma_A = 0.03$ centimeter and $\sigma_B = 0.04$ centimeter. One A piece and one B piece are randomly assembled and laid end to end to form a length about 6 centimeters long. If the assembly is to fit certain quality control standards, it must be between 5.91 and 6.09 centimeters long. What percentage of such assemblies will fail to fall within these limits?

Stop reading and try to solve the problem.

We set as a subgoal that we must determine the distribution function for the sum of two random variables, A and B. We know from background statistical knowledge that, if A and B are normally distributed random variables, then $A + B$ will be a normally distributed random variable with a mean equal to the sum of the means of the component random variables and a variance equal to the sum of the variances. Thus, $\mu_{A+B} = 6$ centimeters and $\sigma_{A+B} = \sqrt{0.03^2 + 0.04^2} = \sqrt{0.05^2} = 0.05$. Having achieved the subgoal of determining the distribution function for the random variable $A + B$, we now set the second and final subgoal to be to determine what percentage of the distribution lies outside a region of 0.09 on either side of the mean. This subgoal can be determined from a table of the normal distribution, provided we know how many standard deviations is represented by 0.09 centimeter. To determine this amount we simply divide 0.09 by 0.05 to get 1.8 standard deviation units, telling us that we are asking for the percentage of cases falling in the two tails of a normal distribution 1.8 standard deviation units out from the mean. Looking up the value $z = 1.8$ in a table of the normal distribution gives the figure of 0.036 in one tail or 7.2 percent in both tails. Once again, the solution of the problem proceeds from our setting a series of one or more subgoals that, taken together, constitute the solution to the entire problem.

> Determine one way in which the random variable z might have been formed, where the moment generating function of z is
>
> $$M_z(\theta) = \frac{e^{4\theta + (\theta^2/2)}}{(1 - 2\theta)^{3/2}}$$

Stop reading and try to solve the problem.

In contrast to earlier problems, in this one the goal is specified but the givens are not, and we must determine some set of givens such that the goal—namely, the moment generating function—can be derived as a consequence. The obvious way to check out any hypothesized set of givens is to use the method of contradiction. In addition, we should use the method of working backward from the goal expression, since this is a unique starting point in the problem. The most relevant piece of background information is that the moment generating function of a sum of random variables is the product of the moment generating functions of each component random variable. Stop reading and try again to solve the problem, if you did not before.

By examining a table of such moment generating functions, we can quickly exclude the possibility that the moment generating function of z is itself a random variable with a simple standard distribution function.

The next simplest hypothesis would be that z is the sum of two random variables, each of which has a simple familiar distribution function. This being the case, we should work backward from the goal expression by factoring it into two components, each of which is a moment generating function for a familiar distribution function. The most obvious split of the goal moment generating function is probably to multiply the numerator times the reciprocal of the denominator. It turns out that the numerator is the moment generating function for a normal distribution with mean 4 and standard deviation 1, and the reciprocal of the denominator is the moment generating function for a random variable of the χ^2 distribution on 3 degrees of freedom, which means that z is the sum of these two random variables.

Had this particular factorization not worked, there are a number of other simple factorizations of the moment generating function that might have been matched for form against our table of moment generating functions for familiar distributions. In many ways, the solution of the problem is terribly simple. We can ask, "How can we not start with the goal expression, which is the only given in the problem other than implicit given information?" Indeed, if we start to manipulate the goal expression and know the relevant background information about the moment generating functions, it is difficult to see how we can fail to solve the problem. Nevertheless, many people do fail to solve this problem and other equally simple problems, because they have no idea what to do. In many cases, they are genuinely deficient in important background information, but those who knew they had to work backward from the goal expression in the present problem would likely look up in books the relevant information about moment generating functions that was needed in order to solve the problem. Those who have a thorough knowledge of the specific subject matter probably need to have no conscious understanding of general problem-solving methods in order to solve this and many other problems. However, those who are learning the specific subject matter will be aided in this learning by a thorough knowledge of general problem-solving methods, which suggest what types of information are needed in order to solve problems.

Again, in the problem with the two pieces A and B that are joined end-to-end to form a new combined piece that must fall within certain tolerance limits, students might lack the specific background information about the distribution function of the sum of two normally distributed random variables. However, having clearly defined the subgoal of determining such a distribution function in order to determine the percentage of cases that lie in its tails, it is likely that students would look for the directly relevant piece of information they lacked.

Formulas for getting certain information from other information often automatically provide you with a set of subgoals — namely, that of determining the values of the various components of these formulas. Thus, if you have enough specific background information to know the appropriate general formulas, you can often substitute that information for an understanding of general problem-solving methods in those cases where you know some general formula that encompasses all the aspects of the problem. However, if no such formula exists or if you do not know it, understanding general problem-solving techniques can be quite crucial in devising an adequate plan to solve the problem.

COMBINATORIAL ANALYSIS

> How many ways can a set of contestants consisting of four men, three women, two boys, and three girls be selected from an audience consisting of eight men, nine women, six boys, and six girls?

Stop reading and try to solve the problem.

To solve the problem, we might set a series of subgoals; that is, we might determine how many ways there are to pick first the men alone, then the women alone, then the boys alone, and then the girls alone. Let us call the solutions to these four subproblems N_1, N_2, N_3, and N_4. Stop reading and try again to solve the problem, if you did not before.

The number of ways to pick an entire set of contestants is simply the product $N_1 \cdot N_2 \cdot N_3 \cdot N_4$. Each of the subgoals is a simple combinations problem (unordered sets obtained by sampling without replacement). Thus, the total number of ways is simply

$$\left(\frac{8 \cdot 7 \cdot 6 \cdot 5}{1 \cdot 2 \cdot 3 \cdot 4}\right)\left(\frac{9 \cdot 8 \cdot 7}{1 \cdot 2 \cdot 3}\right)\left(\frac{6 \cdot 5}{1 \cdot 2}\right)\left(\frac{6 \cdot 5 \cdot 4}{1 \cdot 2 \cdot 3}\right)$$

John and Fred agree to play a tennis match, with the winner to be the person who first wins two sets in a row or a total of three sets. Find the number of ways the match can occur.

Stop reading and try to solve the problem.

The most straightforward way to determine the number of ways the match can occur is to construct a tree diagram, marking all terminals of the tree where either one of the conditions is first satisfied and stopping the growth of the tree from that point on. The tree has two branches at each node—namely, A wins or B wins.

Alternatively, we can determine the answer without explicitly constructing the tree, by the following line of reasoning. We first make certain inferences from the information given in the problem—namely, that the match cannot end before two sets have been played and must end after a maximum of five sets have been played (since out of five sets one player must win at least three sets). Having made these inferences, the problem of determining the number of ways the match can occur can be reduced to a set of four subproblems—namely, we must determine how many ways the match can end after two sets, three sets, four sets, or five sets. Stop reading and try again to solve the problem, if you did not before.

Clearly, there are only two ways the match can end after two sets—namely, A wins both sets or B wins both sets. There are also only two ways the match can end after three sets: A wins the first set and B wins the next two, or B wins the first set and A wins the next two. Now we might note that, in general, at each level of the tree, after the second, there will be exactly two terminal nodes and two nonterminal nodes under the rule that the winning player must win two sets in a row. Thus, at every node prior to the last, there will be exactly two terminal nodes and two nonterminal nodes. At the last node, there will be four terminal nodes, since, by the three-set rule, all nodes must be terminal once five sets have been played. Thus, there are two terminal nodes after two sets, two terminal nodes after three sets, two terminal nodes after four sets, and four terminal nodes after five sets, or 10 terminal nodes in all (and so 10 ways the match can occur).

The principal general problem-solving methods used in solving the problem were inference and the subgoal method. Note that although the specialized method of explictly constructing a tree diagram will also solve this problem without the need for using the more general subgoal method, the subgoal method in combination with certain inferences generalizes easily to problems in which to construct an explicit tree would be extremely laborious.

John pays a quarter to play a simple coin-flipping game against a gambling casino. The quarter entitles him to play a maximum of five coin flips against the house. John wins $1 every time he calls the coin correctly (head or tails) and loses $1 every time he calls the coin incorrectly. John begins with $3 and will stop playing whenever he loses his entire stake or wins $3 (that is, has a total of $6). Of course, he must quit after playing a maximum of five coin flips. Find the number of ways the playing can occur.

Stop reading and try to solve the problem.

A first general problem-solving method we might use is to note the similarity to the previous problem. This similarity leads to the conjecture that we could solve the problem either by constructing an explicit tree diagram or by making certain inferences about the tree diagram and then breaking up the problem into subproblems to determine how many ways the playing can occur, stopping after N coin flips, for all $N \leq 5$. Stop reading and try again to solve the problem, if you did not before.

One simple inference is that John must play for at least three coin flips, since, at worst, he will loose $1 on each coin flip, and he has $3 to play with. Since he can play at most for five coin flips for his original quarter, we know that we can break the problem into three subproblems—namely, to determine how many ways the playing will stop after three, four, and five coin flips. Clearly, there are exactly two ways the playing will stop after three coin flips (two terminal nodes), leaving six nonterminal nodes after three coin flips. Another relevant inference is that it is impossible to be ahead or behind by an even number of dollars after any odd number of coin flips (such as three flips). Thus, it is impossible for John to be either even or ahead or behind by $2 after three flips. Hence, the six nonterminal nodes must all involve different sequences of winnings and losings that total either +$1 or −$1, and, by symmetry, there must be three of each type. From this we can conclude that there are no terminal nodes after four coin flips and 12 nonterminal nodes. The 12 nonterminal nodes at level

4 lead to 24 nodes at level 5, all of which, by definition, must be terminal. Thus, there are precisely 26 different ways the playing can occur.

Binomial theorem. Prove that

$$(a + b)^n = \sum_{r=0}^{n} \binom{n}{r} a^{n-r} \cdot b^r$$

where $\binom{n}{r} = \dfrac{n!}{r!(n-r)!}$

The basic problem-solving method to be used is mathematical induction, which we have already noted involves a combination of the general problem-solving methods of special case (proving the theorem true for $n = 1$) and the subgoal method (dividing the proof of the theorem into two parts: proving it true for $n = 1$ and showing that, if it is true for n, the theorem is true for $n + 1$).

Stop reading and try to solve the problem.

The theorem is trivially true for $n = 1$, so the crux of the proof consists of assuming that the theorem holds for $(a + b)^n$ and proving it is true for $(a + b)^{n+1}$. To prove this lemma, we assume the theorem is true for n and multiply both sides of the equation by $(a + b)$. This operation yields the term $(a + b)^{n+1}$ on the left side of the equation, as desired. The right side of the equation will clearly involve exactly $n + 2$ terms of the form $a^{n-r} \cdot b^r$, where r goes from zero to $n + 1$, as desired. This is obvious by inspection. What remains is to prove that the coefficients of each $a^{n-r}b^r$ term have the form $\binom{n+1}{r}$. Except for the terms a^{n+1} and b^{n+1}, which arise only once in the multiplication of $(a+b) \sum_{r=0}^{n} \binom{n}{r} a^{n-r}b^r$, every other $a^{n-r} \cdot b^r$ term arises in two places. The term in the product that contains b^r is obtained from

$$b\left[\binom{n}{r-1}a^{n-r+1}b^{r-1}\right] + a\left[\binom{n}{r}a^{n-r}b^r\right] = \binom{n}{r-1}a^{n-r+1}b^r + \binom{n}{r}a^{n-r+1}b^r$$

$$= \left[\binom{n}{r-1} + \binom{n}{r}\right]a^{n-r+1}b^r$$

All that remains is to show that $\left[\binom{n}{r-1} + \binom{n}{r}\right] = \binom{n+1}{r}$. This remaining subgoal is trivially established by algebraic combination of the two fractions, and the theorem is proved.

Besides using the method of mathematical induction, which we have noted is an application of two general problem-solving methods, our implementation of the proof involved breaking the problem into two parts: first, determining that there were the correct number and type of terms $a^{n-r+1}b^r$ on the right side of the equation and, second, determining that the coefficients of each such term were of the proper form, namely, $\binom{n+1}{r}$. Achieving the final subgoal of showing that $\left[\binom{n}{r-1}\right.$ $+ \left.\binom{n}{r}\right] = \binom{n+1}{r}$ could be said to involve hill climbing on a two-dimensional evaluation function consisting of the number of terms in the numerator and the denominator of the coefficient of $a^{n-r+1}b^r$. In the goal, the coefficient of $a^{n-r+1}b^r$ is a simple fraction consisting of one factorial in the numerator divided by two factorials in the denominator. The expression $\left[\binom{n}{r-1} + \binom{n}{r}\right]$ clearly involves two separate factorial fractions that must be combined into a single factorial fraction with two factorials in the denominator and one in the numerator. By analogy to similar problems, we first express both fractions in terms of the same denominator by multiplying numerator and denominator of each fraction by the appropriate numbers. Then we add the numerators, putting them over the common denominator, and factor the numerator to obtain a coefficient of $a^{n-r+1}b^r$ of the desired form. Of course, most problem solvers engaged in solving problems of this type will have practiced the latter sequence of operations so well in hundreds of preceding problems that they will hardly need to think of applying any general problem-solving method in order to implement the algebraic solution.

The evaluation function used in defining the initial breakup into sub-problems (determining whether there were the right number of $a^n b^r$ terms and determining whether the coefficients matched) comes straight from characterization of the right side of the goal expression, namely, $\sum_{r=0}^{n+1} \binom{n+1}{r} a^{n-r}b^r$. After multiplying $(a+b) \sum_{r=0}^{n} \binom{n}{r} a^{n-r}b^r$, we obtain $2(n+1)$ terms, and these terms must be reduced to $n+2$ terms of the proper form, in order to achieve the goal. This reduction can be subdivided into two parts: first we achieve the right number of terms having the proper $a^n b^r$ components and then we determine if the coefficients of these terms match the desired coefficients in the goal expression. Note that in order to define the subgoals, it is not necessary to explicitly define any single numerical or vectored-valued evaluation function. All that is necessary is that we have a more or less explicit awareness of some of the dimensions on which the goal expression differs from the given expression and define subgoals on the basis that they match the goal expression on more dimensions than the given expression.

NUMBER THEORY

Prove that if $(2^n - 1)$ is a prime number, then n is a prime number.

This problem and its proof were given to me by Al Stevens. Stop reading and try to solve the problem.

A good general problem-solving method to apply initially is the method of contradiction. This method is suggested by the existence of two simple alternatives for n: either it is prime or it is not prime. If it is not prime, it is expressible as the produce of two integer factors, neither of which equals unity. If n being not prime in conjunction with $2^n - 1$ being prime can be shown to yield the contradictory conclusion that $2^n - 1$ is not prime, then the original theorem will be established. Stop reading and try again to solve the problem, if you did not do so before.

To implement the method we must show n is not prime implies that $2^n - 1$ is not prime. If n is not prime, then $2^n - 1 = 2^{jk} - 1$, where j and $k \geq 2$. Under these circumstances $2^{jk} - 1$ can be factored into

$$(2^j - 1) \left[2^{(k-1)j} + 2^{(k-2)j} + \cdots + 2^{(k-k+1)j} + 2^{(k-k)j} \right]$$

This latter is established by simple division of $(2^j - 1)$ into $(2^{jk} - 1)$. Thus, $2^n - 1$ is not prime, contradicting the given information.

About the only general problem-solving method I can suggest that might give you the idea of trying to use the factor $(2^j - 1)$ would be general experience with problems involving similar expressions — namely, those of the form $a^n - b^m$. Of course, there are not too many obvious factors to try to use other than either $(2^j - 1)$ or $(2^k - 1)$, either of which will do. Thus, once you decide to use the method of contradiction, the rest of the problem is relatively straightforward.

MODERN ALGEBRA

Given that the positive integers are well ordered (each nonempty subset of the integers contains exactly one smallest integer), prove that there is no integer between 0 and 1.

Stop reading and try to solve the problem.

Use the method of contradiction. Now try again to solve the problem, if you did not do so before.

Assume that there are one or more integers between 0 and 1. By the well-ordering property, there is in this nonempty set of integers between 0 and 1 some least integer, m, for which $0 < m < 1$. Multiplying both sides of these inequalities by the number m, we have $0 < m^2 < m$. Thus, m^2 must be another integer in the class of integers between 0 and 1 and, furthermore, $m^2 < m$, which contradicts the assumption that m was the least integer between 0 and 1. Since the contradiction was reached by assuming the existence of integers between 0 and 1, this implies that there is no integer between 0 and 1.

> Given a set of elements, G, with a binary operation, *, defined over G such that G is a group. The definition of the group is a set of elements with a binary operation such that all of the following four properties hold. (1) Closure: for all a, b, and c in G, $a*b = d$ is a member of G. (2) Associativity: for all a, b, and c in G, $(a*b)*c = a*(b*c)$. (3) Left identity: for all a in G, there exists an e in G such that $e*a = a$. (4) Left inverse: for all a in G, there exists an a^{-1} in G such that $a^{-1}*a = e$. For such a system prove the following theorem: Unique left inverse: for all a in G, there exists a *unique* a^{-1} in G such that $a^{-1}*a = e$.

Stop reading and try to prove the above theorem.

Whenever we encounter a uniqueness proof, the method of contradiction is immediately suggested. That is, we should assume that there exist two different left inverses (a^{-1} and a_1^{-1}) such that $a^{-1}*a = e$ and $a_1^{-1}*a = e$, and attempt to show that $a^{-1} = a_1^{-1}$ (contradicting the assumption that a^{-1} and a_1^{-1} are different). Stop reading and try again to solve the problem.

It is trivial to conclude that, if $a^{-1}*a = e$ and $a_1^{-1}*a = e$, then $a^{-1}*a = a_1^{-1}*a$. However, it may appear somewhat difficult to peel off the identical a's from the right-hand side of the equation (since we have not already proved any right cancellation law). Therefore, we might set as a subgoal (lemma) to prove the right cancellation law—namely, that for all b in G, $b*a = c*a$ implies that $b = c$. However, this law is certainly not easier to prove than the unique left inverse theorem itself. Thus, this subgoal appears unlikely to be useful in the present problem. However, there is another subgoal (conjectured lemma) that is easier to establish and thus facilitates solution of the present problem. This lemma does allow us in essence to peel off the a from the equation $a^{-1}*a = a_1^{-1}*a$. Try to conjecture this lemma (subgoal) and then prove it, if you have not done so already.

The useful lemma (subgoal) is that the left inverse of an element in a group is the same as the right inverse — namely, $a^{-1}*a = e$ implies that $a*a^{-1} = e$. Clearly, if this lemma were true, it would permit us to multiply both sides of the equation $a^{-1}*a = a_1^{-1}*a$ on the right by the quantity a^{-1} and change the identical a's into e's on both sides of the equation. Since we have special given information regarding e's, they might be easier to peel off than a's. Stop reading and try to prove the lemma that the left inverse equals the right inverse.

Proof of this lemma involves use of the inference method; that is, we simply perform substitution operations on the quantity $a*a^{-1}$ to attempt to show that $a*a^{-1} = e$. The exact proof of this lemma follows.

$$a*a^{-1} = a*(e*a^{-1}) = a*(a^{-1}*a)*a^{-1} = a*(a^{-1}*(a*a^{-1}))$$

$$= (a*a^{-1})*(a*a^{-1})$$

Let $a*a^{-1} = b$, then $b = b*b$. There exists b^{-1} in G such that $b^{-1}*b = e$.

$$b = e*b = (b^{-1}*b)*b = b^{-1}*(b*b) = b^{-1}*b = e$$

Thus, $b = e$ and $a*a^{-1} = e$ Q.E.D.

Now stop reading and try to prove the rest of the theorem, if you have not done so already.

The first lemma does not quite permit us to prove the theorem in a straightforward way, since what we obtain is an expression of the form $a^{-1}*e = a_1^{-1}*e$, and we are not yet justified in dropping the e's from both sides of the equation. We have been given the left identity property but not the right identity property. Therefore, it is necessary to set a second subgoal (lemma) of proving the right identity property for a group — namely, that $e*a = a$ implies that $a*e = a$ for all a in G. Stop reading and prove this second lemma and then continue to prove the rest of the theorem, if you have not done so already.

The proof of the second lemma is quite trivial again by the inference method and is given below:

$$a*e = a*(a^{-1}*a) = (a*a^{-1})*a = e*a = a$$ Q.E.D.

Given the above two lemmas, the proof of the original theorem concerning the uniqueness of the left inverse is quite trivial and is given below:

$$a^{-1}*a = e   \text{and}   a_1^{-1}*a = e$$

$$(a^{-1}*a)*a^{-1} = (a_1^{-1}*a)*a^{-1}$$

$$a^{-1}*(a*a^{-1}) = a_1^{-1}*(a*a^{-1})$$

But, by the first lemma, $a*a^{-1} = e$. Therefore, $a^{-1}*e = a_1^{-1}*e$, and, by the second lemma, $a^{-1} = a_1^{-1}$ (Q.E.D.).

MECHANICS

What constant force will cause a mass of 3 kilograms to achieve the speed of 30 meters per second in 6 seconds starting from rest? Relevant background information is Newton's second law: $f = ma$, where $a = dv/dt = d^2x/dt^2$. Also relevant is some very elementary knowledge of calculus.

Stop reading and try to solve the problem.

Since we know a formula for the goal quantity (force), the first step is to work backward from the goal and write down the known formula for force—namely, $f = ma$. Since we know the mass (m), we immediately define as a subgoal the determination of the acceleration (a). Stop reading and try again to solve the problem, if you did not do so before.

By definition, $a = dv/dt$ and $a = d^2x/dt^2$. We choose to work with the former formula, since the given information involves velocities (v) and not positions (x). Since we do not know dv/dt, but only certain values of the velocity at the beginning and end of the motion ($v_1 = 0$ and $v_2 = 30$ meters per second), we set another subgoal of transforming the equation $a = dv/dt$ into an equation relating a to the known quantities v_1 and v_2. Stop reading and try again to solve the problem, if you did not before.

Elementary knowledge of calculus tells us that this solution is achieved by use of the integration operation, yielding $a\,dt = v\,dv$, $\int a\,dt = \int v\,dv$, and since a is a constant over time, $at = v + C$. The constant of integration $C = v_1 = 0$, since $v = 0$ at $t = 0$. Thus, we have the formula $a = (v_2 - v_1)/t$, which implies that $a = (30 - 0)/6 = 5$ meters/sec^2. Having achieved the subgoal of determining the acceleration, the rest is simple. Since the mass equals 3 kilograms, force equals $3 \cdot 5 = 15$ newtons.

The principal problem-solving methods used were (a) that of working backward to determine the principal subgoal in the problem, namely, determining the (constant) acceleration, and (b) that of hill climbing in the determination of acceleration, using an evaluation function concerned with how close to given quantities the quantities were on the right side of the equation. By this latter evaluation function, we choose dv/dt over d^2x/dt^2, since the former is at least in some way concerned with a quantity (velocity) that is known at some points. By contrast, the latter expression is concerned with position, about which nothing is known at all in the given information. Furthermore, we choose to transform dv/dt, which is a statement about the derivative of velocity, into a statement about v's at certain points in time, since the latter are directly known from the given information and the former is not.

HEAT

A calorimeter contains 500 grams of water and 300 grams of ice, all at a temperature of 0° C. A 1,000 gram mass of an unknown substance is taken from a furnace where its temperature was 240° C and is dropped immediately into the calorimeter. As a consequence, all the ice is just melted with the temperature of the water remaining at 0°. What would be the final temperature of the water had the mass of the unknown substance been 2,000 grams? Neglect heat loss from the calorimeter and the heat capacity of the calorimeter. The relevant background information for solving this problem consists of the following. The heat of fusion of water equals 80 cal/gm, which means that 80 calories of heat must be supplied to convert 1 gram of ice at 0° C to 1 gram of water at 0° C. Materials are considered to have an approximately constant specific heat capacity (c) over modest ranges of temperatures (consider the ranges discussed in the present experiment to be "modest"). When a body changes temperature, the heat gained or lost equals the mass of the body times the specific heat capacity times the difference in temperature (in degrees C). Finally, under the conditions of calorimeter experiments, the law of conservation of heat holds – namely, heat lost equals heat gained.

Stop reading and try to solve the problem.

It is desirable to introduce efficient symbolic notation to represent the unknown quantities in the present problem. Let t_2 be the temperature of the system (the 2,000 gram substance and the water in the calorimeter after the 2,000 gram substance has been dropped into the water and allowed to reach equilibrium). Let c_s be the specific heat capacity of the unknown substance. The goal is to solve for t_2, but to do so, we must set a subgoal. What is it? Stop reading and try to solve the problem, if you have not done so already.

The obvious subgoal is to determine the specific heat capacity of the unknown substance. This subgoal can evidently be achieved by using the results from the first experiment, where a 1,000 gram mass of the substance was just sufficient to melt 300 grams of ice without changing its temperature. Using the formula that heat lost equals heat gained, we know that $1000 \cdot c_s(240 - 0) = 80 \cdot 300$ or $c_s = 0.1$. Stop reading and solve the rest of the problem, if you have not done so already.

Having solved for the specific heat capacity of the unknown substance, it is now possible to apply the heat-lost-equals-heat-gained formula to the results of the second experiment in order to derive the final temperature of the system after the second experiment. The calculation is as follows:

$$(2000)\,(0.1)\,(240 - t_2) = 500(t_2 - 0) + 300(t_2 - 0) + (80)\,(300)$$

This simple linear equation in one unknown is trivially solved to yield $t_2 = 24°$ C, which is the final temperature of the system.

Principal general problem-solving methods used in the present problem were to label unknown quantities and to set a subgoal. Also, in a problem of this type presented in a physics book, you would have to generate the relevant background information, since it would not be stated explicitly in the problem.

ELECTRICITY

Derive a formula for the electric-field intensity, E, established by a charge distributed uniformly along an infinitely long line with a linear charge density λ. The important background information includes the following: The magnitude of the electric field produced by a point charge of magnitude q at a distance r from the point charge is $E = q/4\pi\epsilon_0 r^2$, where ϵ_0 is a known universal constant that is dependent upon the measuring units, $\lambda = dq/dl$, where l represents position along the line. The direction of the electric-field vector is radially out from the point charge. The electric field produced at a point by a set of point charges is equal to the vector sum of the electric field produced by all component point charges at that point. Also relevant is some knowledge of elementary trigonometry, vectors, and calculus.

Stop reading and try to solve the problem.

The most relevant general problem-solving method to the solution of this problem is for us to define subgoals (break up the problem into parts). What is the first subgoal, we might consider? Stop reading and try again to solve the problem, if you did not before.

Although the problem asks us to describe the entire electric field produced by the line (at an infinity of points in space), we know by analogy to similar problems that this statement means we must derive a formula for the electric field at some arbitrary point in space. Thus, the problem is simplified by considering the electric field at only a single (variable) point in space. Furthermore, symmetry indicates that the only relevant information is the distance of the point from the line, represented by h in Fig. 11-2. Clearly, the electric field set up by a charge distributed along a straight line must have cylindrical symmetry (be equal at all points at the same distance h from the line), since there is nothing different about the given information for any such point. Stop reading and try again to solve the problem, if you did not do so before.

Another useful general problem-solving technique would be to draw a diagram representing the important information in the problem.

The problem can be broken into parts by defining another subgoal. What subgoal might this be? Stop reading and try again to solve the problem.

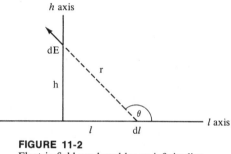

FIGURE 11-2
Electric field produced by an infinite line
with linear charge density λ.

Since we know from a physical assumption that the electric field at a point is equal to the sum of the contributions of the electric field produced by all charges, it is relevant to attempt to determine the individual contribution to the electric field at a point h distant from the line due to any little piece of charge along the line. Consider the electric field produced at the point by the amount of charge present along an infinitesimally small segment of the line dl. The charge in this segment is $dq = \lambda\, dl$. Stop reading and try again to solve the problem by first solving the subgoal, if you have not done so already.

The contribution to the electric field (dE) produced by the charge $\lambda\, dl$ in an infinitesimally small segment of a line (dl) is given by the formula (which we know from background information) $dE = \lambda\, dl/4\pi\epsilon_0 r^2$. Note that dE should be a vector quantity, and we have only obtained an expression for the magnitude of the vector. It is also necessary to state the direction of the vector. According to background information, this direction is evidently radially out from the point charge at dl, as shown in Fig. 11-2. Having achieved the first subgoal, our next subgoal is to combine the contributions to the field from all segments dl along the infinitely long line. This combining will evidently involve an integration, since the segments are infinitesimally small, rather than a summation where the contributions to the field are finite in number. Stop reading and try to solve the rest of the problem, if you have not done so already.

In attempting to combine separate contributions of each dl along the line, it is necessary to note that the directions of the vectors dE produced by each dl are different. Thus, we cannot simply integrate the magnitudes of these vectors with respect to l from minus infinity to plus infinity. Instead, we must resolve each vector into l and h components and integrate each separately with respect to l from minus infinity to plus infinity. Thus, the next subgoal is to resolve the electric field produced by each dl into two components. Elementary trigonometry applied to the previous formula for the electric field produced by dl yields the following components:

$$dE_l = dE \cos \theta = \left(\frac{\lambda \, dl}{4\pi\epsilon_0 r^2} \right)\left(\frac{-l}{r} \right) = \frac{-\lambda l \, dl}{4\pi\epsilon_0 r^3} = \frac{-\lambda l \, dl}{4\pi\epsilon_0 (h^2 + l^2)^{3/2}}$$

$$dE_h = dE \sin \theta = \left(\frac{\lambda \, dl}{4\pi\epsilon_0 r^2} \right)\left(\frac{h}{r} \right) = \frac{\lambda h \, dl}{4\pi\epsilon_0 r^3} = \frac{\lambda h \, dl}{4\pi\epsilon_0 (h^2 + l^2)^{3/2}}$$

Note that in the final expressions for dE_l and dE_h we substituted $(h^2 + l^2)^{1/2}$ for r because we are intending to integrate with respect to l, and r is a function of l. Thus, we must express r in terms of l. This sort of manipulation to eliminate unnecessary terms by expressing them in other necessary terms is a form of hill climbing on an evaluation function concerned with the number of unknown terms. Stop reading and try to solve the rest of the problem, if you have not done so already.

Of course, all that remains now is to actually perform the two integrations to determine the E_l and E_h components of the field at the point h distant from the line. This is shown in the work below:

$$E_l = -\int_{-\infty}^{\infty} \frac{\lambda l \, dl}{4\pi\epsilon_0 (h^2 + l^2)^{3/2}} = \frac{-\lambda}{4\pi\epsilon_0}\left[-\frac{1}{(h^2 + l^2)^{1/2}} \right]_{-\infty}^{\infty}$$

$$= \frac{-\lambda}{4\pi\epsilon_0}\left[-\lim_{l \to \infty}\left(-\frac{1}{(h^2 + l^2)^{1/2}} \right) + \lim_{l \to -\infty}\left(\frac{1}{(h^2 + l^2)^{1/2}} \right) \right]$$

$$= \frac{-\lambda}{4\pi\epsilon_0}[0 + 0] = 0$$

$$E_h = \int_{-\infty}^{\infty} \frac{\lambda h\, dl}{4\pi\epsilon_0 (h^2 + l^2)^{3/2}} = \frac{\lambda h}{4\pi\epsilon_0} \left[\frac{l}{h^2 (h^2 + l^2)^{1/2}} \right]_{-\infty}^{\infty}$$

$$= \frac{\lambda h}{4\pi\epsilon_0} \left[\lim_{l\to\infty} \left(\frac{l}{h^2(h^2 + l^2)^{1/2}} \right) - \lim_{l\to-\infty} \left(\frac{l}{h^2(h^2 + l^2)^{1/2}} \right) \right]$$

$$= \frac{\lambda h}{4\pi\epsilon_0} \left[\frac{1}{h^2} + \frac{1}{h^2} \right] = \frac{2\lambda h}{4\pi\epsilon_0 h^2} = \frac{\lambda}{2\pi\epsilon_0 h}$$

Principal problem-solving methods used in the solution of this electrostatics problem were subgoals, representing information by symbols and diagrams, symmetry (noticing equivalence classes), similarity to previous problems, and perhaps some limited use of hill climbing. We might even contend that the subgoal of computing the contribution to the electric field at a point produced by a small quantity of charge *dq* distributed over a small segment of the line *dl* constituted the solution of a simpler problem and thus was an example of that general problem-solving method, in addition to representing the subgoal method.

ELECTRICAL ENGINEERING

Givens: You have a limited supply of 2-input AND gates and 2-input OR gates to use in constructing a variety of control circuits. A 2-input AND or OR gate has two input wires and one output wire. All input levels and output levels are either 0 or 1 (binary digital-logic circuits). A 2-input AND gate has a 1 on the output wire, if and only if both input wires are at the 1 level. A 2-input OR gate has a 1 on the output wire, if and only if either one or both of its inputs is at the 1 level. In constructing control circuits, it is important to know that you may connect the same (source) wire to many different input wires of many different gates. Also, the output wire of one gate may be connected to one or more input wires of one or more gates in chains and even loops. In particular, the output of a gate may be connected to one of its own inputs. However, you may not connect two outputs. A somewhat related restriction is that you must not connect two wires to the same input wire, whenever doing so would complete an undesirable circuit between the two wires. In the present problem, assume that all such circuits are undesirable and do not connect two wires to the same input wire of a gate.

Goal: In the part of the circuit you are now constructing, there are 6 input wires and 15 output wires, each of which can be at the 0 level or the 1 level. Only 15 patterns of 0's and 1's will ever occur on the 6 input wires. Your task is to choose the set of 15 input patterns and construct a decoding circuit, using the minimum number of 2-input AND or OR gates, such that when any one of these 15 input patterns occurs, one and only one of the 15 output wires will be at the 1 level (the rest being at 0). Naturally, a different output wire should be at the 1 level for each of the 15 different input patterns.

Stop reading and try to solve the problem.
The first step here, as in any problem, is to explore the problem, deriving whatever conclusions can be derived easily. For example, with 6 binary inputs, there are 64 possible input patterns, only 15 of which are being used. Four binary input wires would suffice to present 15 different input patterns, so there must be some advantage in using more input wires. It would be a good guess that it simplifies the decoding circuit and minimizes the number of gates to employ 6 input wires, rather than 4. Along the same line, it would be reasonable to conjecture that the problem would be essentially solved if we knew which 15 input patterns to use.

If we were inclined toward number theory, we might inquire about the properties of the numbers 6 or 15. In this particular problem, such an inquiry could yield an immediate idea for the correct solution, especially if the problem solver were already sufficiently familiar with the use of AND and OR gates in circuit problems. However, let us not follow up this specific approach to the problem now. You can go back to consider this approach, after we have gone through more straightforward and more general methods.

Another thing we might derive is the conclusion that the solution of the problem must require at least 15 gates, one for each different output wire. If the problem can be solved with 15 gates, this number must be the minimum. Of course, we do not know yet whether a greater number of gates than 15 will be required. Stop reading and try again to solve the problem, if you did not do so before.

So much for deriving quick conclusions. If you are inexperienced in using AND and OR gates in circuit problems, you will probably want to spend some time thinking about their properties and using them in a more or less random way, unrelated to the problem. You are probably somewhat familiar with *and* and *or* as used in logical expressions, but psychologically that is not quite the same as using AND and OR gates as transformations (operators) in circuit-design problems.

Near the beginning of your work on the problem, you should develop useful representations of concepts in it. Vector notation for the six-bit binary input patterns will probably aid your thinking—for example, 110100 or 001000. Some sort of spatial representation of the two different kinds of gates (labeled boxes), and the input and output wires (lines) might also be helpful to you. Stop reading and try again to solve the problem, if you did not before.

These somewhat ponderous preliminaries to real work on the problem may seem completely unnecessary to some, but those for whom the preliminaries are unnecessary are either lucky in solving this particular problem or else are consciously or unconsciously accomplishing these preliminaries very quickly in their brains. Once a person becomes skilled at problem solving, these preliminaries, which take several paragraphs to explain, can be accomplished in seconds in the head.

Having accomplished the preliminaries of fully understanding the problem, deriving quick conclusions, and developing some useful verbal and spatial representations, it is time to see if some solution to the problem just pops into your head, probably because it is analogous to similar problems you have solved in the past. If nothing comes to mind, you might try more actively to think about whether you have solved similar problems in the past and what specific or general methods you used then. Let us assume that this is a failure; you never encountered a circuit problem before in your life or, in any event, you have not remembered anything that seems useful from previous problems.

What next? You might try breaking the problem into parts (subproblems or subgoals). For example, you could note that three input wires can have eight different input patterns, which means that perhaps the problem could be broken down into two subgoals: that of mapping eight input patterns on wires 1, 2, and 3 onto eight of the output wires, and that of mapping seven input patterns on wires 4, 5, and 6 onto the remaining seven output wires. It seems a trifle inelegant to have been given 15, instead of 16, codes, but in a real-world problem, nothing guarantees this kind of elegance. Of course, this is a made-up problem, and it is more elegant than this. Ignoring the question of inelegance, we might spend some time trying this approach based on the subgoal method. However, it happens in this case that the analysis into subproblems is not helpful. There is a power in the combinations across the two sets of three input lines that is being lost by this analysis into subproblems. I have not thought of any other analysis of this problem into subproblems that is helpful, either. Thus, the subgoal method is a complete bust on this problem, but if you tried this method you would be making a relatively intelligent error. What other general problem-solving method might you use? Stop reading and try again to solve the problem, if you have not already.

Many other general problem-solving methods could be tried, but the one that really cracks the problem open in a systematic, straightforward, though somewhat time-consuming, manner is to solve simpler problems. There are a large number of simpler problems. You can start as simple as you wish and work your way up through more complicated problems, and hope the general principle of the solution to the original problem becomes clear. Stop reading and try again to solve the problem, if you did not before.

A good subproblem to start with would be five input patterns on four input wires, to be decoded onto five output wires. There are only 16 possible binary input patterns on four wires, and you are to select five of the 16 to achieve a circuit using the smallest number of gates. Presumably, in working on this subproblem, you learn a number of principles that will be useful in solving the original problem. For example, you learn to focus on the input wires, which are at the 1 level in any particular input pattern, because, without any invertors, it is only the 1-level inputs that can be used to turn on the correct output. Also, presumably you realize (if you did not already) that some circuits will turn on the correct output but also turn on some incorrect output wires, in violation of the requirements of the problem. This type of difficulty might incline you against selecting binary input codes that had too many 1's in them.

Note that we avoided choosing a simpler problem that had no more input patterns than it had input lines, because such a problem permits the trivial one-to-one solution that obviously will not work in the original problem and will give no insights into the original problem. Thus, we may already have realized that many or all of the input patterns must have more than a single 1 in them. In the present simpler problem, the combination of not wanting too many 1's and wanting more than a single 1 in many or all of the input patterns essentially forces us to use different combinations of two 1's as inputs to AND gates—for example, 1100, 1010, 1001, 0110, 0101. At this point, we might see that this type of solution generalizes directly to the original problem, or we might solve another, slightly more difficult problem before seeing that the solution generalizes to the original problem. Stop reading and try again to solve the problem, if you did not before.

Let us back up a little. Maybe we never thought explicitly of the principle that too many 1's in an input pattern are no good. Nevertheless, we would be apt to obtain the solution to the simpler problem because the range of possible solutions is so much reduced. To be sure, there are a lot of different combinations of 16 patterns taken five at a time, and we must also describe the decoding circuit for any five we select. However, the number of logically different *classes* of potential solutions is much smaller than this. Without trying to enumerate all of the logically different classes of sets of five input patterns, we can indicate the nature of the features used to define these classes according to whether the pattern used consists of all 1's (1111); three 1's (such as 1110); two 1's (such as 1100); one 1 (such as 0100); all 0's (0000); or whether the same wire is at the 1 level in all of the five patterns, four of the five, three of the five, and so on. If you use the method of classificatory trial and error (being systematic about noting the features of the types of solutions you have considered and rejected), it should not take too long to hit upon the optimal solution to the simpler problem. Stop reading and try again to solve the problem, if you have not done so already.

The solution to the original problem is to choose the following set of 15 input patterns: 110000, 101000, 100100, 100010, 100001, 011000, 010100, 010010, 010001, 001100, 001010, 001001, 000110, 000101, 000011.

The decoding circuit uses just 15 AND gates with input lines 1 and 2 connected to the first AND gate, input lines 1 and 3 to the second, input lines 1 and 4 to the third, and so on, up to input lines 5 and 6 to the 15th.

What if you chose so trivial a simpler problem that no useful insights were obtained? An example would be three input patterns on three input wires to be decoded onto three output wires. If you avoided the trivial one-to-one solution, you could still learn the necessary principles from this simpler problem. If you did not avoid the trivial solution, you could then pose a somewhat more complex problem, continuing this process until a problem was posed that was simple enough to solve easily but hard enough to involve some of the important principles of the solution to the original problem.

What if your judgment regarding simpler problems was faulty, and a harder problem was selected? For example, you might think that eight input patterns on four input wires was a simpler problem than the original problem (15 input patterns on six wires), but it is not. Furthermore, a solution to this problem will tend to lead you away from the optimal solution to the original problem. You must honestly face the fact that this is a potential trap that often accompanies the use of the simpler-problem method. Your criteria for judging the simplicity of a problem are vitally important for the success of the method. If you know this, explicitly specify criteria for problem simplicity within any class of problems, and continually question these criteria for problem simplicity when the supposedly simpler problem proves difficult, then you can avoid being trapped by the method.

What if, in working on the original problem or some simpler problem, you develop mental sets (unconscious assumptions) about the solution to the problem that are wrong and that prevent you from obtaining the necessary ideas for solving the problem? This often happens, especially when a person does not have a habit of continually trying to specify the methods being used and the assumptions being made.

For example, in the present problem, we can develop the working hypothesis that somehow the six input wires should be considered in three groups of two wires each. This sort of crudely formulated working hypothesis could be very helpful, if it were correct. In this problem it is not correct, and it can be distinctly deleterious for getting the necessary ideas.

As usual, an ounce of prevention is worth a pound of cure. If you are careful to note the working assumptions you make, it will be easy to question those assumptions and think of other ideas (working assumptions) that violate them. However, sometimes even the most analytical problem solvers make unconscious working assumptions and then find themselves going around in circles—that is, repeatedly trying out the same incorrect solutions within a limited set that does not contain the correct solution. If you are aware of this possibility, then you can try to characterize your implicit assumptions and make one or more contrary assumptions.

COMPUTER PROGRAMMING

Computer programming problems provide particularly good examples of subgoals, the representation of information (naming), inference (representation of implicit information), analogy, and special case.

Computer programming problems frequently involve the solution of one or more mathematical problems, such as deriving an algorithm for the solution of an equation, in addition to the definition of a sequence of instructions to achieve the solution of the problem by the computer. This problem solving already provides an example of the subgoal method, with one subgoal being the mathematical solution of one or more problems and a second subgoal being the representation of this solution in a programming language.

Another basic application of the subgoal method to virtually all computer programming problems is the division of the problem into three parts: input, computation, and output. In addition to the input of the program itself, most programming problems require the values of certain variables to be input to the machine. The computer must be told from what source to expect the input (cards, magnetic tape, paper tape, or whatever) and the format of the input (alphabetic, alpha-numeric, numerical, two-column fields, three-column fields, and so on). In addition, the computer must be told where to store this data and what names to give to the various subsets of input data. These instructions represent further subsubgoals of the input phase of the programming problem. Similarly, the computer must be told what values from what arrays to output, on what output medium (printer, cards, magnetic tape, paper tape, and so on), the output format, and alphanumeric headings for various portions of the output.

The computational portion of any large program must frequently be divided further into subgoals, the solution to each of these subgoals being called a subroutine. For example, it may be that a portion of the computation involved in a computer program is to find the value of a function such as $y = f(x) = a \log x + b \sin x + c \sqrt{x}$. Without con-sidering exactly where in the program this subroutine will be used, a computer programmer might write up a program for the computation of this function and give that portion of the program (subroutine) a name, so that it might be called at any point during the execution of the main program. The computation of values of functions constitutes but one relatively trivial example of the application of the subgoal method to computer programming problems. Other examples of sub-routines include random-number generators, shuffling programs, find-ing the maximum value in an array of values, ordering or ranking a set of numbers, and searching for a particular alphanumeric label. Fre-quently, the programmer knows a number of subroutines that will be required to solve a computer programming problem, programs to achieve these subroutines can be developed relatively independently of one another and of the main program. Because of the large inde-pendence that can be achieved in writing a computer program to achieve various subgoals, it is possible for a team of programmers to divide the work of writing a large program. However, it is frequently necessary for certain common naming conventions to be observed and for the writer of the main program to specify to the writers of each of the component subroutines the form and location of the input to their subroutines and the desired form and location for the output from the subroutines.

The importance of giving names to all the important concepts in a programming problem is so obviously forced upon any programmer by the necessity of representing every important aspect of the problem in a computer program that it bears no extensive discussion.

The frequent need to represent implicit information in the solution of programming problems is almost, but not quite, as obvious as the necessity of naming important concepts. For example, assume that one required subroutine is to sample randomly from a set without replacement. We might achieve this by simulating card shuffling in the computer. To simulate card shuffling in a computer it is necessary to represent explicitly what we know implicitly to be involved in shuffling a deck of cards. Shuffling a deck of cards involves first making a single partition at some random point near the middle of the deck in order to divide the set into two subsets (two interval subsets). The cut might easily be achieved in a computer by picking a random number between 0 and 9 and adding this to a number that is 5 less than half of the number of cards in the deck. Having simulated the cut, we are now faced with simulating the actual shuffle. What is evidently involved in this shuffle is that the top card from one of the two subsets is inserted at some random point within the top few cards of the other subset and the second card is inserted randomly a few cards below the first card, and so on. This can be easily simulated on a computer by picking a random number between 0 and 2 for the number of cards from the other subset to intervene between any two adjacent cards from the first subset. This is not necessarily the best shuffling routine, but it vividly illustrates the process of explicitly representing implicit information as a component to the solution of a programming problem.

Analogy is widely used in the solution of programming problems in the relatively trivial sense that whenever a problem can be identified as being essentially identical to a problem for which a program already exists, a programmer will obtain that program from some library and incorporate it into his own program to solve that portion of the problem. This method plays an important role in solving computer programming problems, but its use is so widely understood that it hardly deserves much comment here.

Finally, the method of special case often plays an important role in the solution of computer programming problems. Programming problems frequently involve doing certain computational jobs over and over again for some multidimensional array of values or vectors of values as the input. It simplifies the problem greatly to first write a program to solve the problem in a special case and then extend this solution to the entire multidimensional array. Very frequently, this method amounts to little more than getting a subroutine for doing a particular job (such as computing the value of a function) and then embedding that subroutine within a set of control loops that iterate the subroutine through all of the values in an input matrix and output the results of the computation into the proper places in an output matrix. The method of special case is sometimes equivalent to the subgoal method.

References

Bartlett, F. *Thinking: An experiment and social study.* New York: Basic Books, 1958.

Chessin, P. L. Problem for solution. *American Mathematical Monthly,* 1954, *61,* 258–59.

Duncker, K. On problem solving. *Psychological Monographs,* 1945, *58* (5, Whole No. 270).

Feller, W. *An introduction to probability theory and its applications.* (3rd ed.) Vol. 1. New York: John Wiley & Sons, 1957.

Newell, A., Shaw, J. C., & Simon, H. A. The processes of creative thinking. In H. E. Gruber, G. Terrell, & M. Wertheimer (Eds.), *Contemporary Approaches to Creative Thinking.* New York: Atherton Press, 1962. Pp. 63–110.

Polya, G. *How to solve it.* Garden City, N.Y.: Doubleday & Company, 1957.

Polya, G. *Mathematical discovery.* Vol. 1. *On understanding, learning, and teaching problem solving.* New York: John Wiley & Sons, 1962.

Simon, H. A., & Newell, A. Human problem solving. *American Psychologist,* 1971, *26,* 145–159.

Index

A CATALOG OF SELECTED
DOVER BOOKS
IN ALL FIELDS OF INTEREST

A CATALOG OF SELECTED DOVER
BOOKS IN ALL FIELDS OF INTEREST

CONCERNING THE SPIRITUAL IN ART, Wassily Kandinsky. Pioneering work by father of abstract art. Thoughts on color theory, nature of art. Analysis of earlier masters. 12 illustrations. 80pp. of text. 5⅜ × 8½.   23411-8 Pa. $3.95

ANIMALS: 1,419 Copyright-Free Illustrations of Mammals, Birds, Fish, Insects, etc., Jim Harter (ed.). Clear wood engravings present, in extremely lifelike poses, over 1,000 species of animals. One of the most extensive pictorial sourcebooks of its kind. Captions. Index. 284pp. 9 × 12.   23766-4 Pa. $12.95

CELTIC ART: The Methods of Construction, George Bain. Simple geometric techniques for making Celtic interlacements, spirals, Kells-type initials, animals, humans, etc. Over 500 illustrations. 160pp. 9 × 12. (USO)   22923-8 Pa. $9.95

AN ATLAS OF ANATOMY FOR ARTISTS, Fritz Schider. Most thorough reference work on art anatomy in the world. Hundreds of illustrations, including selections from works by Vesalius, Leonardo, Goya, Ingres, Michelangelo, others. 593 illustrations. 192pp. 7⅛ × 10¼.   20241-0 Pa. $9.95

CELTIC HAND STROKE-BY-STROKE (Irish Half-Uncial from "The Book of Kells"): An Arthur Baker Calligraphy Manual, Arthur Baker. Complete guide to creating each letter of the alphabet in distinctive Celtic manner. Covers hand position, strokes, pens, inks, paper, more. Illustrated. 48pp. 8¼ × 11.
24336-2 Pa. $3.95

EASY ORIGAMI, John Montroll. Charming collection of 32 projects (hat, cup, pelican, piano, swan, many more) specially designed for the novice origami hobbyist. Clearly illustrated easy-to-follow instructions insure that even beginning papercrafters will achieve successful results. 48pp. 8¼ × 11.   27298-2 Pa. $2.95

THE COMPLETE BOOK OF BIRDHOUSE CONSTRUCTION FOR WOOD-WORKERS, Scott D. Campbell. Detailed instructions, illustrations, tables. Also data on bird habitat and instinct patterns. Bibliography. 3 tables. 63 illustrations in 15 figures. 48pp. 5¼ × 8½.   24407-5 Pa. $1.95

BLOOMINGDALE'S ILLUSTRATED 1886 CATALOG: Fashions, Dry Goods and Housewares, Bloomingdale Brothers. Famed merchants' extremely rare catalog depicting about 1,700 products: clothing, housewares, firearms, dry goods, jewelry, more. Invaluable for dating, identifying vintage items. Also, copyright-free graphics for artists, designers. Co-published with Henry Ford Museum & Greenfield Village. 160pp. 8¼ × 11.   25780-0 Pa. $9.95

HISTORIC COSTUME IN PICTURES, Braun & Schneider. Over 1,450 costumed figures in clearly detailed engravings—from dawn of civilization to end of 19th century. Captions. Many folk costumes. 256pp. 8⅜ × 11¾.   23150-X Pa. $11.95

CATALOG OF DOVER BOOKS

STICKLEY CRAFTSMAN FURNITURE CATALOGS, Gustav Stickley and L. & J. G. Stickley. Beautiful, functional furniture in two authentic catalogs from 1910. 594 illustrations, including 277 photos, show settles, rockers, armchairs, reclining chairs, bookcases, desks, tables. 183pp. 6½ × 9¼. 23838-5 Pa. $9.95

AMERICAN LOCOMOTIVES IN HISTORIC PHOTOGRAPHS: 1858 to 1949, Ron Ziel (ed.). A rare collection of 126 meticulously detailed official photographs, called "builder portraits," of American locomotives that majestically chronicle the rise of steam locomotive power in America. Introduction. Detailed captions. xi + 129pp. 9 × 12. 27393-8 Pa. $12.95

AMERICA'S LIGHTHOUSES: An Illustrated History, Francis Ross Holland, Jr. Delightfully written, profusely illustrated fact-filled survey of over 200 American lighthouses since 1716. History, anecdotes, technological advances, more. 240pp. 8 × 10¾. 25576-X Pa. $11.95

TOWARDS A NEW ARCHITECTURE, Le Corbusier. Pioneering manifesto by founder of "International School." Technical and aesthetic theories, views of industry, economics, relation of form to function, "mass-production split" and much more. Profusely illustrated. 320pp. 6⅛ × 9¼. (USO) 25023-7 Pa. $9.95

HOW THE OTHER HALF LIVES, Jacob Riis. Famous journalistic record, exposing poverty and degradation of New York slums around 1900, by major social reformer. 100 striking and influential photographs. 233pp. 10 × 7⅞. 22012-5 Pa $10.95

FRUIT KEY AND TWIG KEY TO TREES AND SHRUBS, William M. Harlow. One of the handiest and most widely used identification aids. Fruit key covers 120 deciduous and evergreen species; twig key 160 deciduous species. Easily used. Over 300 photographs. 126pp. 5⅜ × 8½. 20511-8 Pa. $3.95

COMMON BIRD SONGS, Dr. Donald J. Borror. Songs of 60 most common U.S. birds: robins, sparrows, cardinals, bluejays, finches, more—arranged in order of increasing complexity. Up to 9 variations of songs of each species. Cassette and manual 99911-4 $8.95

ORCHIDS AS HOUSE PLANTS, Rebecca Tyson Northen. Grow cattleyas and many other kinds of orchids—in a window, in a case, or under artificial light. 63 illustrations. 148pp. 5⅜ × 8½. 23261-1 Pa. $4.95

MONSTER MAZES, Dave Phillips. Masterful mazes at four levels of difficulty. Avoid deadly perils and evil creatures to find magical treasures. Solutions for all 32 exciting illustrated puzzles. 48pp. 8¼ × 11. 26005-4 Pa. $2.95

MOZART'S DON GIOVANNI (DOVER OPERA LIBRETTO SERIES), Wolfgang Amadeus Mozart. Introduced and translated by Ellen H. Bleiler. Standard Italian libretto, with complete English translation. Convenient and thoroughly portable—an ideal companion for reading along with a recording or the performance itself. Introduction. List of characters. Plot summary. 121pp. 5¼ × 8½. 24944-1 Pa. $2.95

TECHNICAL MANUAL AND DICTIONARY OF CLASSICAL BALLET, Gail Grant. Defines, explains, comments on steps, movements, poses and concepts. 15-page pictorial section. Basic book for student, viewer. 127pp. 5⅜ × 8½. 21843-0 Pa. $4.95

BRASS INSTRUMENTS: Their History and Development, Anthony Baines. Authoritative, updated survey of the evolution of trumpets, trombones, bugles, cornets, French horns, tubas and other brass wind instruments. Over 140 illustrations and 48 music examples. Corrected and updated by author. New preface. Bibliography. 320pp. 5⅜ × 8½. 27574-4 Pa. $9.95

HOLLYWOOD GLAMOR PORTRAITS, John Kobal (ed.). 145 photos from 1926–49. Harlow, Gable, Bogart, Bacall; 94 stars in all. Full background on photographers, technical aspects. 160pp. 8⅜ × 11¼. 23352-9 Pa. $11.95

MAX AND MORITZ, Wilhelm Busch. Great humor classic in both German and English. Also 10 other works: "Cat and Mouse," "Plisch and Plumm," etc. 216pp. 5⅜ × 8½. 20181-3 Pa. $5.95

THE RAVEN AND OTHER FAVORITE POEMS, Edgar Allan Poe. Over 40 of the author's most memorable poems: "The Bells," "Ulalume," "Israfel," "To Helen," "The Conqueror Worm," "Eldorado," "Annabel Lee," many more. Alphabetic lists of titles and first lines. 64pp. 5³/₁₆ × 8¼. 26685-0 Pa. $1.00

SEVEN SCIENCE FICTION NOVELS, H. G. Wells. The standard collection of the great novels. Complete, unabridged. First Men in the Moon, Island of Dr. Moreau, War of the Worlds, Food of the Gods, Invisible Man, Time Machine, In the Days of the Comet. Total of 1,015pp. 5⅜ × 8½. (USO) 20264-X Clothbd. $29.95

AMULETS AND SUPERSTITIONS, E. A. Wallis Budge. Comprehensive discourse on origin, powers of amulets in many ancient cultures: Arab, Persian, Babylonian, Assyrian, Egyptian, Gnostic, Hebrew, Phoenician, Syriac, etc. Covers cross, swastika, crucifix, seals, rings, stones, etc. 584pp. 5⅜ × 8½. 23573-4 Pa. $12.95

RUSSIAN STORIES/PYCCKNE PACCKA3bl: A Dual-Language Book, edited by Gleb Struve. Twelve tales by such masters as Chekhov, Tolstoy, Dostoevsky, Pushkin, others. Excellent word-for-word English translations on facing pages, plus teaching and study aids, Russian/English vocabulary, biographical/critical introductions, more. 416pp. 5⅜ × 8½. 26244-8 Pa. $8.95

PHILADELPHIA THEN AND NOW: 60 Sites Photographed in the Past and Present, Kenneth Finkel and Susan Oyama. Rare photographs of City Hall, Logan Square, Independence Hall, Betsy Ross House, other landmarks juxtaposed with contemporary views. Captures changing face of historic city. Introduction. Captions. 128pp. 8¼ × 11. 25790-8 Pa. $9.95

AIA ARCHITECTURAL GUIDE TO NASSAU AND SUFFOLK COUNTIES, LONG ISLAND, The American Institute of Architects, Long Island Chapter, and the Society for the Preservation of Long Island Antiquities. Comprehensive, well-researched and generously illustrated volume brings to life over three centuries of Long Island's great architectural heritage. More than 240 photographs with authoritative, extensively detailed captions. 176pp. 8¼ × 11. 26946-9 Pa. $14.95

NORTH AMERICAN INDIAN LIFE: Customs and Traditions of 23 Tribes, Elsie Clews Parsons (ed.). 27 fictionalized essays by noted anthropologists examine religion, customs, government, additional facets of life among the Winnebago, Crow, Zuni, Eskimo, other tribes. 480pp. 6⅛ × 9¼. 27377-6 Pa. $10.95

CATALOG OF DOVER BOOKS

FRANK LLOYD WRIGHT'S HOLLYHOCK HOUSE, Donald Hoffmann. Lavishly illustrated, carefully documented study of one of Wright's most controversial residential designs. Over 120 photographs, floor plans, elevations, etc. Detailed perceptive text by noted Wright scholar. Index. 128pp. 9¼ × 10¾.
27133-1 Pa. $11.95

THE MALE AND FEMALE FIGURE IN MOTION: 60 Classic Photographic Sequences, Eadweard Muybridge. 60 true-action photographs of men and women walking, running, climbing, bending, turning, etc., reproduced from rare 19th-century masterpiece. vi + 121pp. 9 × 12. 24745-7 Pa. $10.95

1001 QUESTIONS ANSWERED ABOUT THE SEASHORE, N. J. Berrill and Jacquelyn Berrill. Queries answered about dolphins, sea snails, sponges, starfish, fishes, shore birds, many others. Covers appearance, breeding, growth, feeding, much more. 305pp. 5¼ × 8¼. 23366-9 Pa. $7.95

GUIDE TO OWL WATCHING IN NORTH AMERICA, Donald S. Heintzelman. Superb guide offers complete data and descriptions of 19 species: barn owl, screech owl, snowy owl, many more. Expert coverage of owl-watching equipment, conservation, migrations and invasions, etc. Guide to observing sites. 84 illustrations. xiii + 193pp. 5⅜ × 8½. 27344-X Pa. $8.95

MEDICINAL AND OTHER USES OF NORTH AMERICAN PLANTS: A Historical Survey with Special Reference to the Eastern Indian Tribes, Charlotte Erichsen-Brown. Chronological historical citations document 500 years of usage of plants, trees, shrubs native to eastern Canada, northeastern U.S. Also complete identifying information. 343 illustrations. 544pp. 6½ × 9¼. 25951-X Pa. $12.95

STORYBOOK MAZES, Dave Phillips. 23 stories and mazes on two-page spreads: Wizard of Oz, Treasure Island, Robin Hood, etc. Solutions. 64pp. 8¼ × 11.
23628-5 Pa. $2.95

NEGRO FOLK MUSIC, U.S.A., Harold Courlander. Noted folklorist's scholarly yet readable analysis of rich and varied musical tradition. Includes authentic versions of over 40 folk songs. Valuable bibliography and discography. xi + 324pp. 5⅜ × 8½. 27350-4 Pa. $7.95

MOVIE-STAR PORTRAITS OF THE FORTIES, John Kobal (ed.). 163 glamor, studio photos of 106 stars of the 1940s: Rita Hayworth, Ava Gardner, Marlon Brando, Clark Gable, many more. 176pp. 8⅜ × 11¼. 23546-7 Pa. $11.95

BENCHLEY LOST AND FOUND, Robert Benchley. Finest humor from early 30s, about pet peeves, child psychologists, post office and others. Mostly unavailable elsewhere. 73 illustrations by Peter Arno and others. 183pp. 5⅜ × 8½.
22410-4 Pa. $5.95

YEKL and THE IMPORTED BRIDEGROOM AND OTHER STORIES OF YIDDISH NEW YORK, Abraham Cahan. Film Hester Street based on Yekl (1896). Novel, other stories among first about Jewish immigrants on N.Y.'s East Side. 240pp. 5⅜ × 8½. 22427-9 Pa. $6.95

SELECTED POEMS, Walt Whitman. Generous sampling from *Leaves of Grass.* Twenty-four poems include "I Hear America Singing," "Song of the Open Road," "I Sing the Body Electric," "When Lilacs Last in the Dooryard Bloom'd," "O Captain! My Captain!"—all reprinted from an authoritative edition. Lists of titles and first lines. 128pp. 5³⁄₁₆ × 8¼. 26878-0 Pa. $1.00

THE BEST TALES OF HOFFMANN, E. T. A. Hoffmann. 10 of Hoffmann's most important stories: "Nutcracker and the King of Mice," "The Golden Flowerpot," etc. 458pp. 5⅜ × 8½. 21793-0 Pa. $8.95

FROM FETISH TO GOD IN ANCIENT EGYPT, E. A. Wallis Budge. Rich detailed survey of Egyptian conception of "God" and gods, magic, cult of animals, Osiris, more. Also, superb English translations of hymns and legends. 240 illustrations. 545pp. 5⅜ × 8½. 25803-3 Pa. $11.95

FRENCH STORIES/CONTES FRANÇAIS: A Dual-Language Book, Wallace Fowlie. Ten stories by French masters, Voltaire to Camus: "Micromegas" by Voltaire; "The Atheist's Mass" by Balzac; "Minuet" by de Maupassant; "The Guest" by Camus, six more. Excellent English translations on facing pages. Also French-English vocabulary list, exercises, more. 352pp. 5⅜ × 8½. 26443-2 Pa. $8.95

CHICAGO AT THE TURN OF THE CENTURY IN PHOTOGRAPHS: 122 Historic Views from the Collections of the Chicago Historical Society, Larry A. Viskochil. Rare large-format prints offer detailed views of City Hall, State Street, the Loop, Hull House, Union Station, many other landmarks, circa 1904–1913. Introduction. Captions. Maps. 144pp. 9⅜ × 12¼. 24656-6 Pa. $12.95

OLD BROOKLYN IN EARLY PHOTOGRAPHS, 1865–1929, William Lee Younger. Luna Park, Gravesend race track, construction of Grand Army Plaza, moving of Hotel Brighton, etc. 157 previously unpublished photographs. 165pp. 8⅞ × 11¾. 23587-4 Pa. $13.95

THE MYTHS OF THE NORTH AMERICAN INDIANS, Lewis Spence. Rich anthology of the myths and legends of the Algonquins, Iroquois, Pawnees and Sioux, prefaced by an extensive historical and ethnological commentary. 36 illustrations. 480pp. 5⅜ × 8½. 25967-6 Pa. $8.95

AN ENCYCLOPEDIA OF BATTLES: Accounts of Over 1,560 Battles from 1479 B.C. to the Present, David Eggenberger. Essential details of every major battle in recorded history from the first battle of Megiddo in 1479 B.C. to Grenada in 1984. List of Battle Maps. New Appendix covering the years 1967–1984. Index. 99 illustrations. 544pp. 6½ × 9¼. 24913-1 Pa. $14.95

SAILING ALONE AROUND THE WORLD, Captain Joshua Slocum. First man to sail around the world, alone, in small boat. One of great feats of seamanship told in delightful manner. 67 illustrations. 294pp. 5⅜ × 8½. 20326-3 Pa. $5.95

ANARCHISM AND OTHER ESSAYS, Emma Goldman. Powerful, penetrating, prophetic essays on direct action, role of minorities, prison reform, puritan hypocrisy, violence, etc. 271pp. 5⅜ × 8½. 22484-8 Pa. $5.95

MYTHS OF THE HINDUS AND BUDDHISTS, Ananda K. Coomaraswamy and Sister Nivedita. Great stories of the epics; deeds of Krishna, Shiva, taken from puranas, Vedas, folk tales; etc. 32 illustrations. 400pp. 5⅜ × 8½. 21759-0 Pa. $9.95

BEYOND PSYCHOLOGY, Otto Rank. Fear of death, desire of immortality, nature of sexuality, social organization, creativity, according to Rankian system. 291pp. 5⅜ × 8½. 20485-5 Pa. $8.95

A THEOLOGICO-POLITICAL TREATISE, Benedict Spinoza. Also contains unfinished Political Treatise. Great classic on religious liberty, theory of government on common consent. R. Elwes translation. Total of 421pp. 5⅜ × 8½. 20249-6 Pa. $8.95

MY BONDAGE AND MY FREEDOM, Frederick Douglass. Born a slave, Douglass became outspoken force in antislavery movement. The best of Douglass' autobiographies. Graphic description of slave life. 464pp. 5⅜ × 8½. 22457-0 Pa. $8.95

FOLLOWING THE EQUATOR: A Journey Around the World, Mark Twain. Fascinating humorous account of 1897 voyage to Hawaii, Australia, India, New Zealand, etc. Ironic, bemused reports on peoples, customs, climate, flora and fauna, politics, much more. 197 illustrations. 720pp. 5⅜ × 8½. 26113-1 Pa. $15.95

THE PEOPLE CALLED SHAKERS, Edward D. Andrews. Definitive study of Shakers: origins, beliefs, practices, dances, social organization, furniture and crafts, etc. 33 illustrations. 351pp. 5⅜ × 8½. 21081-2 Pa. $8.95

THE MYTHS OF GREECE AND ROME, H. A. Guerber. A classic of mythology, generously illustrated, long prized for its simple, graphic, accurate retelling of the principal myths of Greece and Rome, and for its commentary on their origins and significance. With 64 illustrations by Michelangelo, Raphael, Titian, Rubens, Canova, Bernini and others. 480pp. 5⅜ × 8½. 27584-1 Pa. $9.95

PSYCHOLOGY OF MUSIC, Carl E. Seashore. Classic work discusses music as a medium from psychological viewpoint. Clear treatment of physical acoustics, auditory apparatus, sound perception, development of musical skills, nature of musical feeling, host of other topics. 88 figures. 408pp. 5⅜ × 8½. 21851-1 Pa. $9.95

THE PHILOSOPHY OF HISTORY, Georg W. Hegel. Great classic of Western thought develops concept that history is not chance but rational process, the evolution of freedom. 457pp. 5⅜ × 8½. 20112-0 Pa. $9.95

THE BOOK OF TEA, Kakuzo Okakura. Minor classic of the Orient: entertaining, charming explanation, interpretation of traditional Japanese culture in terms of tea ceremony. 94pp. 5⅜ × 8½. 20070-1 Pa. $3.95

LIFE IN ANCIENT EGYPT, Adolf Erman. Fullest, most thorough, detailed older account with much not in more recent books, domestic life, religion, magic, medicine, commerce, much more. Many illustrations reproduce tomb paintings, carvings, hieroglyphs, etc. 597pp. 5⅜ × 8½. 22632-8 Pa. $10.95

SUNDIALS, Their Theory and Construction, Albert Waugh. Far and away the best, most thorough coverage of ideas, mathematics concerned, types, construction, adjusting anywhere. Simple, nontechnical treatment allows even children to build several of these dials. Over 100 illustrations. 230pp. 5⅜ × 8½. 22947-5 Pa. $7.95

DYNAMICS OF FLUIDS IN POROUS MEDIA, Jacob Bear. For advanced students of ground water hydrology, soil mechanics and physics, drainage and irrigation engineering, and more. 335 illustrations. Exercises, with answers. 784pp. 6⅛ × 9¼. 65675-6 Pa. $19.95

SONGS OF EXPERIENCE: Facsimile Reproduction with 26 Plates in Full Color, William Blake. 26 full-color plates from a rare 1826 edition. Includes "The Tyger," "London," "Holy Thursday," and other poems. Printed text of poems. 48pp. 5¼ × 7. 24636-1 Pa. $4.95

OLD-TIME VIGNETTES IN FULL COLOR, Carol Belanger Grafton (ed.). Over 390 charming, often sentimental illustrations, selected from archives of Victorian graphics—pretty women posing, children playing, food, flowers, kittens and puppies, smiling cherubs, birds and butterflies, much more. All copyright-free. 48pp. 9¼ × 12¼. 27269-9 Pa. $5.95

PERSPECTIVE FOR ARTISTS, Rex Vicat Cole. Depth, perspective of sky and sea, shadows, much more, not usually covered. 391 diagrams, 81 reproductions of drawings and paintings. 279pp. 5⅜ × 8½. 22487-2 Pa. $6.95

DRAWING THE LIVING FIGURE, Joseph Sheppard. Innovative approach to artistic anatomy focuses on specifics of surface anatomy, rather than muscles and bones. Over 170 drawings of live models in front, back and side views, and in widely varying poses. Accompanying diagrams. 177 illustrations. Introduction. Index. 144pp. 8⅜ × 11¼. 26723-7 Pa. $8.95

GOTHIC AND OLD ENGLISH ALPHABETS: 100 Complete Fonts, Dan X. Solo. Add power, elegance to posters, signs, other graphics with 100 stunning copyright-free alphabets: Blackstone, Dolbey, Germania, 97 more—including many lower-case, numerals, punctuation marks. 104pp. 8⅛ × 11. 24695-7 Pa. $8.95

HOW TO DO BEADWORK, Mary White. Fundamental book on craft from simple projects to five-bead chains and woven works. 106 illustrations. 142pp. 5⅜ × 8. 20697-1 Pa. $4.95

THE BOOK OF WOOD CARVING, Charles Marshall Sayers. Finest book for beginners discusses fundamentals and offers 34 designs. "Absolutely first rate . . . well thought out and well executed."—E. J. Tangerman. 118pp. 7¾ × 10⅝. 23654-4 Pa. $5.95

ILLUSTRATED CATALOG OF CIVIL WAR MILITARY GOODS: Union Army Weapons, Insignia, Uniform Accessories, and Other Equipment, Schuyler, Hartley, and Graham. Rare, profusely illustrated 1846 catalog includes Union Army uniform and dress regulations, arms and ammunition, coats, insignia, flags, swords, rifles, etc. 226 illustrations. 160pp. 9 × 12. 24939-5 Pa. $10.95

WOMEN'S FASHIONS OF THE EARLY 1900s: An Unabridged Republication of "New York Fashions, 1909," National Cloak & Suit Co. Rare catalog of mail-order fashions documents women's and children's clothing styles shortly after the turn of the century. Captions offer full descriptions, prices. Invaluable resource for fashion, costume historians. Approximately 725 illustrations. 128pp. 8⅜ × 11¼. 27276-1 Pa. $11.95

THE 1912 AND 1915 GUSTAV STICKLEY FURNITURE CATALOGS, Gustav Stickley. With over 200 detailed illustrations and descriptions, these two catalogs are essential reading and reference materials and identification guides for Stickley furniture. Captions cite materials, dimensions and prices. 112pp. 6½ × 9¼. 26676-1 Pa. $9.95

EARLY AMERICAN LOCOMOTIVES, John H. White, Jr. Finest locomotive engravings from early 19th century: historical (1804–74), main-line (after 1870), special, foreign, etc. 147 plates. 142pp. 11⅜ × 8¼. 22772-3 Pa. $10.95

THE TALL SHIPS OF TODAY IN PHOTOGRAPHS, Frank O. Braynard. Lavishly illustrated tribute to nearly 100 majestic contemporary sailing vessels: Amerigo Vespucci, Clearwater, Constitution, Eagle, Mayflower, Sea Cloud, Victory, many more. Authoritative captions provide statistics, background on each ship. 190 black-and-white photographs and illustrations. Introduction. 128pp. 8⅞ × 11¾. 27163-3 Pa. $13.95

EARLY NINETEENTH-CENTURY CRAFTS AND TRADES, Peter Stockham (ed.). Extremely rare 1807 volume describes to youngsters the crafts and trades of the day: brickmaker, weaver, dressmaker, bookbinder, ropemaker, saddler, many more. Quaint prose, charming illustrations for each craft. 20 black-and-white line illustrations. 192pp. 4⅜ × 6. 27293-1 Pa. $4.95

VICTORIAN FASHIONS AND COSTUMES FROM HARPER'S BAZAR, 1867-1898, Stella Blum (ed.). Day costumes, evening wear, sports clothes, shoes, hats, other accessories in over 1,000 detailed engravings. 320pp. 9⅜ × 12¼. 22990-4 Pa. $13.95

GUSTAV STICKLEY, THE CRAFTSMAN, Mary Ann Smith. Superb study surveys broad scope of Stickley's achievement, especially in architecture. Design philosophy, rise and fall of the Craftsman empire, descriptions and floor plans for many Craftsman houses, more. 86 black-and-white halftones. 31 line illustrations. Introduction. 208pp. 6½ × 9¼. 27210-9 Pa. $9.95

THE LONG ISLAND RAIL ROAD IN EARLY PHOTOGRAPHS, Ron Ziel. Over 220 rare photos, informative text document origin (1844) and development of rail service on Long Island. Vintage views of early trains, locomotives, stations, passengers, crews, much more. Captions. 8⅞ × 11¾. 26301-0 Pa. $13.95

THE BOOK OF OLD SHIPS: From Egyptian Galleys to Clipper Ships, Henry B. Culver. Superb, authoritative history of sailing vessels, with 80 magnificent line illustrations. Galley, bark, caravel, longship, whaler, many more. Detailed, informative text on each vessel by noted naval historian. Introduction. 256pp. 5⅜ × 8½. 27332-6 Pa. $6.95

TEN BOOKS ON ARCHITECTURE, Vitruvius. The most important book ever written on architecture. Early Roman aesthetics, technology, classical orders, site selection, all other aspects. Morgan translation. 331pp. 5⅜ × 8½. 20645-9 Pa. $8.95

THE HUMAN FIGURE IN MOTION, Eadweard Muybridge. More than 4,500 stopped-action photos, in action series, showing undraped men, women, children jumping, lying down, throwing, sitting, wrestling, carrying, etc. 390pp. 7⅞ × 10⅝. 20204-6 Clothbd. $24.95

TREES OF THE EASTERN AND CENTRAL UNITED STATES AND CANADA, William M. Harlow. Best one-volume guide to 140 trees. Full descriptions, woodlore, range, etc. Over 600 illustrations. Handy size. 288pp. 4½ × 6⅜. 20395-6 Pa. $5.95

SONGS OF WESTERN BIRDS, Dr. Donald J. Borror. Complete song and call repertoire of 60 western species, including flycatchers, juncoes, cactus wrens, many more—includes fully illustrated booklet. Cassette and manual 99913-0 $8.95

GROWING AND USING HERBS AND SPICES, Milo Miloradovich. Versatile handbook provides all the information needed for cultivation and use of all the herbs and spices available in North America. 4 illustrations. Index. Glossary. 236pp. 5⅜ × 8½. 25058-X Pa. $6.95

BIG BOOK OF MAZES AND LABYRINTHS, Walter Shepherd. 50 mazes and labyrinths in all—classical, solid, ripple, and more—in one great volume. Perfect inexpensive puzzler for clever youngsters. Full solutions. 112pp. 8⅛ × 11. 22951-3 Pa. $4.95

CATALOG OF DOVER BOOKS

PIANO TUNING, J. Cree Fischer. Clearest, best book for beginner, amateur. Simple repairs, raising dropped notes, tuning by easy method of flattened fifths. No previous skills needed. 4 illustrations. 201pp. 5⅜ × 8½.   23267-0 Pa. $5.95

A SOURCE BOOK IN THEATRICAL HISTORY, A. M. Nagler. Contemporary observers on acting, directing, make-up, costuming, stage props, machinery, scene design, from Ancient Greece to Chekhov. 611pp. 5⅜ × 8½.   20515-0 Pa. $11.95

THE COMPLETE NONSENSE OF EDWARD LEAR, Edward Lear. All nonsense limericks, zany alphabets, Owl and Pussycat, songs, nonsense botany, etc., illustrated by Lear. Total of 320pp. 5⅜ × 8½. (USO)   20167-8 Pa. $6.95

VICTORIAN PARLOUR POETRY: An Annotated Anthology, Michael R. Turner. 117 gems by Longfellow, Tennyson, Browning, many lesser-known poets. "The Village Blacksmith," "Curfew Must Not Ring Tonight," "Only a Baby Small," dozens more, often difficult to find elsewhere. Index of poets, titles, first lines. xxiii + 325pp. 5⅜ × 8¼.   27044-0 Pa. $8.95

DUBLINERS, James Joyce. Fifteen stories offer vivid, tightly focused observations of the lives of Dublin's poorer classes. At least one, "The Dead," is considered a masterpiece. Reprinted complete and unabridged from standard edition. 160pp. 5³⁄₁₆ × 8¼.   26870-5 Pa. $1.00

THE HAUNTED MONASTERY and THE CHINESE MAZE MURDERS, Robert van Gulik. Two full novels by van Gulik, set in 7th-century China, continue adventures of Judge Dee and his companions. An evil Taoist monastery, seemingly supernatural events; overgrown topiary maze hides strange crimes. 27 illustrations. 328pp. 5⅜ × 8½.   23502-5 Pa. $7.95

THE BOOK OF THE SACRED MAGIC OF ABRAMELIN THE MAGE, translated by S. MacGregor Mathers. Medieval manuscript of ceremonial magic. Basic document in Aleister Crowley, Golden Dawn groups. 268pp. 5⅜ × 8½.
23211-5 Pa. $8.95

NEW RUSSIAN-ENGLISH AND ENGLISH-RUSSIAN DICTIONARY, M. A. O'Brien. This is a remarkably handy Russian dictionary, containing a surprising amount of information, including over 70,000 entries. 366pp. 4½ × 6⅛.
20208-9 Pa. $9.95

HISTORIC HOMES OF THE AMERICAN PRESIDENTS, Second, Revised Edition, Irvin Haas. A traveler's guide to American Presidential homes, most open to the public, depicting and describing homes occupied by every American President from George Washington to George Bush. With visiting hours, admission charges, travel routes. 175 photographs. Index. 160pp. 8¼ × 11. 26751-2 Pa. $10.95

NEW YORK IN THE FORTIES, Andreas Feininger. 162 brilliant photographs by the well-known photographer, formerly with *Life* magazine. Commuters, shoppers, Times Square at night, much else from city at its peak. Captions by John von Hartz. 181pp. 9¼ × 10¾.   23585-8 Pa. $12.95

INDIAN SIGN LANGUAGE, William Tomkins. Over 525 signs developed by Sioux and other tribes. Written instructions and diagrams. Also 290 pictographs. 111pp. 6⅛ × 9¼.   22029-X Pa. $3.50

CATALOG OF DOVER BOOKS

ANATOMY: A Complete Guide for Artists, Joseph Sheppard. A master of figure drawing shows artists how to render human anatomy convincingly. Over 460 illustrations. 224pp. 8⅜ × 11¼. 27279-6 Pa. $10.95

MEDIEVAL CALLIGRAPHY: Its History and Technique, Marc Drogin. Spirited history, comprehensive instruction manual covers 13 styles (ca. 4th century thru 15th). Excellent photographs; directions for duplicating medieval techniques with modern tools. 224pp. 8⅜ × 11¼. 26142-5 Pa. $11.95

DRIED FLOWERS: How to Prepare Them, Sarah Whitlock and Martha Rankin. Complete instructions on how to use silica gel, meal and borax, perlite aggregate, sand and borax, glycerine and water to create attractive permanent flower arrangements. 12 illustrations. 32pp. 5⅜ × 8½. 21802-3 Pa. $1.00

EASY-TO-MAKE BIRD FEEDERS FOR WOODWORKERS, Scott D. Campbell. Detailed, simple-to-use guide for designing, constructing, caring for and using feeders. Text, illustrations for 12 classic and contemporary designs. 96pp. 5⅜ × 8½. 25847-5 Pa. $2.95

OLD-TIME CRAFTS AND TRADES, Peter Stockham. An 1807 book created to teach children about crafts and trades open to them as future careers. It describes in detailed, nontechnical terms 24 different occupations, among them coachmaker, gardener, hairdresser, lacemaker, shoemaker, wheelwright, copper-plate printer, milliner, trunkmaker, merchant and brewer. Finely detailed engravings illustrate each occupation. 192pp. 4⅝ × 6. 27398-9 Pa. $4.95

THE HISTORY OF UNDERCLOTHES, C. Willett Cunnington and Phyllis Cunnington. Fascinating, well-documented survey covering six centuries of English undergarments, enhanced with over 100 illustrations: 12th-century laced-up bodice, footed long drawers (1795), 19th-century bustles, 19th-century corsets for men, Victorian "bust improvers," much more. 272pp. 5⅜ × 8¼. 27124-2 Pa. $9.95

ARTS AND CRAFTS FURNITURE: The Complete Brooks Catalog of 1912, Brooks Manufacturing Co. Photos and detailed descriptions of more than 150 now very collectible furniture designs from the Arts and Crafts movement depict davenports, settees, buffets, desks, tables, chairs, bedsteads, dressers and more, all built of solid, quarter-sawed oak. Invaluable for students and enthusiasts of antiques, Americana and the decorative arts. 80pp. 6½ × 9¼. 27471-3 Pa. $7.95

HOW WE INVENTED THE AIRPLANE: An Illustrated History, Orville Wright. Fascinating firsthand account covers early experiments, construction of planes and motors, first flights, much more. Introduction and commentary by Fred C. Kelly. 76 photographs. 96pp. 8¼ × 11. 25662-6 Pa. $8.95

THE ARTS OF THE SAILOR: Knotting, Splicing and Ropework, Hervey Garrett Smith. Indispensable shipboard reference covers tools, basic knots and useful hitches; handsewing and canvas work, more. Over 100 illustrations. Delightful reading for sea lovers. 256pp. 5⅜ × 8½. 26440-8 Pa. $7.95

FRANK LLOYD WRIGHT'S FALLINGWATER: The House and Its History, Second, Revised Edition, Donald Hoffmann. A total revision—both in text and illustrations—of the standard document on Fallingwater, the boldest, most personal architectural statement of Wright's mature years, updated with valuable new material from the recently opened Frank Lloyd Wright Archives. "Fascinating"—The New York Times. 116 illustrations. 128pp. 9¼ × 10¾. 27430-6 Pa. $10.95

PHOTOGRAPHIC SKETCHBOOK OF THE CIVIL WAR, Alexander Gardner. 100 photos taken on field during the Civil War. Famous shots of Manassas, Harper's Ferry, Lincoln, Richmond, slave pens, etc. 244pp. 10⅝ × 8¼.
22731-6 Pa. $9.95

FIVE ACRES AND INDEPENDENCE, Maurice G. Kains. Great back-to-the-land classic explains basics of self-sufficient farming. The one book to get. 95 illustrations. 397pp. 5⅜ × 8½. 20974-1 Pa. $7.95

SONGS OF EASTERN BIRDS, Dr. Donald J. Borror. Songs and calls of 60 species most common to eastern U.S.: warblers, woodpeckers, flycatchers, thrushes, larks, many more in high-quality recording. Cassette and manual 99912-2 $8.95

A MODERN HERBAL, Margaret Grieve. Much the fullest, most exact, most useful compilation of herbal material. Gigantic alphabetical encyclopedia, from aconite to zedoary, gives botanical information, medical properties, folklore, economic uses, much else. Indispensable to serious reader. 161 illustrations. 888pp. 6½ × 9¼. 2-vol. set. (USO) Vol. I: 22798-7 Pa. $9.95
Vol. II: 22799-5 Pa. $9.95

HIDDEN TREASURE MAZE BOOK, Dave Phillips. Solve 34 challenging mazes accompanied by heroic tales of adventure. Evil dragons, people-eating plants, bloodthirsty giants, many more dangerous adversaries lurk at every twist and turn. 34 mazes, stories, solutions. 48pp. 8¼ × 11. 24566-7 Pa. $2.95

LETTERS OF W. A. MOZART, Wolfgang A. Mozart. Remarkable letters show bawdy wit, humor, imagination, musical insights, contemporary musical world; includes some letters from Leopold Mozart. 276pp. 5⅜ × 8½. 22859-2 Pa. $7.95

BASIC PRINCIPLES OF CLASSICAL BALLET, Agrippina Vaganova. Great Russian theoretician, teacher explains methods for teaching classical ballet. 118 illustrations. 175pp. 5⅜ × 8½. 22036-2 Pa. $4.95

THE JUMPING FROG, Mark Twain. Revenge edition. The original story of The Celebrated Jumping Frog of Calaveras County, a hapless French translation, and Twain's hilarious "retranslation" from the French. 12 illustrations. 66pp. 5⅜ × 8½.
22686-7 Pa. $3.95

BEST REMEMBERED POEMS, Martin Gardner (ed.). The 126 poems in this superb collection of 19th- and 20th-century British and American verse range from Shelley's "To a Skylark" to the impassioned "Renascence" of Edna St. Vincent Millay and to Edward Lear's whimsical "The Owl and the Pussycat." 224pp. 5⅜ × 8½.
27165-X Pa. $4.95

COMPLETE SONNETS, William Shakespeare. Over 150 exquisite poems deal with love, friendship, the tyranny of time, beauty's evanescence, death and other themes in language of remarkable power, precision and beauty. Glossary of archaic terms. 80pp. 5³⁄₁₆ × 8¼. 26686-9 Pa. $1.00

BODIES IN A BOOKSHOP, R. T. Campbell. Challenging mystery of blackmail and murder with ingenious plot and superbly drawn characters. In the best tradition of British suspense fiction. 192pp. 5⅜ × 8½. 24720-1 Pa. $5.95

CATALOG OF DOVER BOOKS

THE WIT AND HUMOR OF OSCAR WILDE, Alvin Redman (ed.). More than 1,000 ripostes, paradoxes, wisecracks: Work is the curse of the drinking classes; I can resist everything except temptation; etc. 258pp. 5⅜ × 8½. 20602-5 Pa. $5.95

SHAKESPEARE LEXICON AND QUOTATION DICTIONARY, Alexander Schmidt. Full definitions, locations, shades of meaning in every word in plays and poems. More than 50,000 exact quotations. 1,485pp. 6½ × 9¼. 2-vol. set.
Vol. I: 22726-X Pa. $16.95
Vol. 2: 22727-8 Pa. $15.95

SELECTED POEMS, Emily Dickinson. Over 100 best-known, best-loved poems by one of America's foremost poets, reprinted from authoritative early editions. No comparable edition at this price. Index of first lines. 64pp. 5³⁄₁₆ × 8¼.
26466-1 Pa. $1.00

CELEBRATED CASES OF JUDGE DEE (DEE GOONG AN), translated by Robert van Gulik. Authentic 18th-century Chinese detective novel; Dee and associates solve three interlocked cases. Led to van Gulik's own stories with same characters. Extensive introduction. 9 illustrations. 237pp. 5⅜ × 8½.
23337-5 Pa. $6.95

THE MALLEUS MALEFICARUM OF KRAMER AND SPRENGER, translated by Montague Summers. Full text of most important witchhunter's "bible," used by both Catholics and Protestants. 278pp. 6⅝ × 10. 22802-9 Pa. $11.95

SPANISH STORIES/CUENTOS ESPAÑOLES: A Dual-Language Book, Angel Flores (ed.). Unique format offers 13 great stories in Spanish by Cervantes, Borges, others. Faithful English translations on facing pages. 352pp. 5⅜ × 8½.
25399-6 Pa. $8.95

THE CHICAGO WORLD'S FAIR OF 1893: A Photographic Record, Stanley Appelbaum (ed.). 128 rare photos show 200 buildings, Beaux-Arts architecture, Midway, original Ferris Wheel, Edison's kinetoscope, more. Architectural emphasis; full text. 116pp. 8¼ × 11. 23990-X Pa. $9.95

OLD QUEENS, N.Y., IN EARLY PHOTOGRAPHS, Vincent F. Seyfried and William Asadorian. Over 160 rare photographs of Maspeth, Jamaica, Jackson Heights, and other areas. Vintage views of DeWitt Clinton mansion, 1939 World's Fair and more. Captions. 192pp. 8⅞ × 11. 26358-4 Pa. $12.95

CAPTURED BY THE INDIANS: 15 Firsthand Accounts, 1750–1870, Frederick Drimmer. Astounding true historical accounts of grisly torture, bloody conflicts, relentless pursuits, miraculous escapes and more, by people who lived to tell the tale. 384pp. 5⅜ × 8½. 24901-8 Pa. $8.95

THE WORLD'S GREAT SPEECHES, Lewis Copeland and Lawrence W. Lamm (eds.). Vast collection of 278 speeches of Greeks to 1970. Powerful and effective models; unique look at history. 842pp. 5⅜ × 8½. 20468-5 Pa. $14.95

THE BOOK OF THE SWORD, Sir Richard F. Burton. Great Victorian scholar/adventurer's eloquent, erudite history of the "queen of weapons"—from prehistory to early Roman Empire. Evolution and development of early swords, variations (sabre, broadsword, cutlass, scimitar, etc.), much more. 336pp. 6⅛ × 9¼. 25434-8 Pa. $8.95

CATALOG OF DOVER BOOKS

AUTOBIOGRAPHY: The Story of My Experiments with Truth, Mohandas K. Gandhi. Boyhood, legal studies, purification, the growth of the Satyagraha (nonviolent protest) movement. Critical, inspiring work of the man responsible for the freedom of India. 480pp. 5⅜ × 8½. (USO) 24593-4 Pa. $8.95

CELTIC MYTHS AND LEGENDS, T. W. Rolleston. Masterful retelling of Irish and Welsh stories and tales. Cuchulain, King Arthur, Deirdre, the Grail, many more. First paperback edition. 58 full-page illustrations. 512pp. 5⅜ × 8½. 26507-2 Pa. $9.95

THE PRINCIPLES OF PSYCHOLOGY, William James. Famous long course complete, unabridged. Stream of thought, time perception, memory, experimental methods; great work decades ahead of its time. 94 figures. 1,391pp. 5⅜×8½. 2-vol. set.
Vol. I: 20381-6 Pa. $12.95
Vol. II: 20382-4 Pa. $12.95

THE WORLD AS WILL AND REPRESENTATION, Arthur Schopenhauer. Definitive English translation of Schopenhauer's life work, correcting more than 1,000 errors, omissions in earlier translations. Translated by E. F. J. Payne. Total of 1,269pp. 5⅜ × 8½. 2-vol. set.
Vol. 1: 21761-2 Pa. $11.95
Vol. 2: 21762-0 Pa. $11.95

MAGIC AND MYSTERY IN TIBET, Madame Alexandra David-Neel. Experiences among lamas, magicians, sages, sorcerers, Bonpa wizards. A true psychic discovery. 32 illustrations. 321pp. 5⅜ × 8½. (USO) 22682-4 Pa. $8.95

THE EGYPTIAN BOOK OF THE DEAD, E. A. Wallis Budge. Complete reproduction of Ani's papyrus, finest ever found. Full hieroglyphic text, interlinear transliteration, word-for-word translation, smooth translation. 533pp. 6½ × 9¼. 21866-X Pa. $9.95

MATHEMATICS FOR THE NONMATHEMATICIAN, Morris Kline. Detailed, college-level treatment of mathematics in cultural and historical context, with numerous exercises. Recommended Reading Lists. Tables. Numerous figures. 641pp. 5⅜ × 8½. 24823-2 Pa. $11.95

THEORY OF WING SECTIONS: Including a Summary of Airfoil Data, Ira H. Abbott and A. E. von Doenhoff. Concise compilation of subsonic aerodynamic characteristics of NACA wing sections, plus description of theory. 350pp. of tables. 693pp. 5⅜ × 8½. 60586-8 Pa. $14.95

THE RIME OF THE ANCIENT MARINER, Gustave Doré, S. T. Coleridge. Doré's finest work; 34 plates capture moods, subtleties of poem. Flawless full-size reproductions printed on facing pages with authoritative text of poem. "Beautiful. Simply beautiful."—Publisher's Weekly. 77pp. 9¼ × 12. 22305-1 Pa. $6.95

NORTH AMERICAN INDIAN DESIGNS FOR ARTISTS AND CRAFTS-PEOPLE, Eva Wilson. Over 360 authentic copyright-free designs adapted from Navajo blankets, Hopi pottery, Sioux buffalo hides, more. Geometrics, symbolic figures, plant and animal motifs, etc. 128pp. 8⅜ × 11. (EUK) 25341-4 Pa. $7.95

SCULPTURE: Principles and Practice, Louis Slobodkin. Step-by-step approach to clay, plaster, metals, stone; classical and modern. 253 drawings, photos. 255pp. 8⅛ × 11. 22960-2 Pa. $10.95

CATALOG OF DOVER BOOKS

THE INFLUENCE OF SEA POWER UPON HISTORY, 1660–1783, A. T. Mahan. Influential classic of naval history and tactics still used as text in war colleges. First paperback edition. 4 maps. 24 battle plans. 640pp. 5⅜ × 8½.
25509-3 Pa. $12.95

THE STORY OF THE TITANIC AS TOLD BY ITS SURVIVORS, Jack Winocour (ed.). What it was really like. Panic, despair, shocking inefficiency, and a little heroism. More thrilling than any fictional account. 26 illustrations. 320pp. 5⅜ × 8½. 20610-6 Pa. $8.95

FAIRY AND FOLK TALES OF THE IRISH PEASANTRY, William Butler Yeats (ed.). Treasury of 64 tales from the twilight world of Celtic myth and legend: "The Soul Cages," "The Kildare Pooka," "King O'Toole and his Goose," many more. Introduction and Notes by W. B. Yeats. 352pp. 5⅜ × 8½. 26941-8 Pa. $8.95

BUDDHIST MAHAYANA TEXTS, E. B. Cowell and Others (eds.). Superb, accurate translations of basic documents in Mahayana Buddhism, highly important in history of religions. The Buddha-karita of Asvaghosha, Larger Sukhavativyuha, more. 448pp. 5⅜ × 8½. 25552-2 Pa. $9.95

ONE TWO THREE . . . INFINITY: Facts and Speculations of Science, George Gamow. Great physicist's fascinating, readable overview of contemporary science: number theory, relativity, fourth dimension, entropy, genes, atomic structure, much more. 128 illustrations. Index. 352pp. 5⅜ × 8½. 25664-2 Pa. $8.95

ENGINEERING IN HISTORY, Richard Shelton Kirby, et al. Broad, nontechnical survey of history's major technological advances: birth of Greek science, industrial revolution, electricity and applied science, 20th-century automation, much more. 181 illustrations. ". . . excellent . . ."—Isis. Bibliography. vii + 530pp. 5⅜ × 8¼.
26412-2 Pa. $14.95

Prices subject to change without notice.

Available at your book dealer or write for free catalog to Dept. GI, Dover Publications, Inc., 31 East 2nd St., Mineola, N.Y. 11501. Dover publishes more than 500 books each year on science, elementary and advanced mathematics, biology, music, art, literary history, social sciences and other areas.